The Management
of Technology

The Management of Technology

Perception and opportunities

Paul Lowe

*Formerly the Head of the Department of Manufacturing
and Engineering Systems
Brunel University*

CHAPMAN & HALL

London · Glasgow · Weinheim · New York · Tokyo · Melbourne · Madras

Published by Chapman & Hall, 2–6 Boundary Row, London, SE1 8HN, UK

Chapman & Hall 2–6 Boundary Row, London SE1 8HN, UK

Blackie Academic & Professional, Wester Cleddens Road,Bishopbriggs, Glasgow G64 2NZ, UK

Chapman & Hall GmbH, Pappelallee 3, 69469 Weinheim, Germany

Chapman & Hall USA, 115 Fifth Avenue, New York, NY 10003, USA

Chapman & Hall Japan, ITP-Japan, Kyowa Building, 3F, 2-2-1 Hirakawacho, Chiyoda-ku, Tokyo 102, Japan

Chapman & Hall Australia, 102 Dodds Street, South Melbourne, Victoria 3205, Australia

Chapman & Hall India, R.Seshadri, 32 Second Main Road, CIT East, Madras 600 035, India

First edition 1995

© 1995 Paul Lowe

11200189

Typeset in 10/12pt Palatino by Mews Photosetting, Beckenham, Kent

Printed in Great Britain by TJ Press (Padstow) Ltd, Padstow, Cornwall

ISBN 0 412 64370 7

A catalogue record for this book is available from the British Library
Library of Congress Catalog Card Number: 95-70858

∞ Printed on permanent acid-free text paper, manufactured in accordance with ANSI/NISO Z39.48-1992 and ANSI/NISO Z39.48-1984 (Permanence of Paper).

To my family and friends

Contents

Contents

Preface

This book is about the nature and the use of technology. The subject and its ramifications are so immense that some boundaries have to be set. The focus here is on the management of technology in industry and on some of its implications for the well-being of an economy.

1. INTENT

In essence, the book is an overview. Its purpose is to give perspective to the management of technology. There is room for the integrated view; to write about technology as an entity and to examine some of its facets. Technology is the focus and related aspects form the context. The book is not a research publication – that field is well served by others. In many ways it is a synthesis, a pulling together of various aspects of technology. It draws from different fields of research to describe relevant techniques of analysis and their practical applications. It also considers, at the macro- and micro-levels, some of the economic and industrial challenges of technology. Specialist aspects are assembled and appropriate references and bibliographies allow the reader to explore particular aspects in greater depth.

Although much of what the book contains reflects the situation and experience of the United Kingdom, its message is universal. Wherever possible, examples from other countries have been used. In any case, technology, as such, has few national boundaries and many of its challenges and problems apply both to industrial and developing countries.

The views expressed in this book are what social scientists call 'positive' and 'normative'. A positive account endeavours to describe what is observed. A normative approach applies value judgements and recommends certain courses of action, which are believed to lead to some desirable goal.

Limited and simple use has been made of mathematics. It is more for illustration than argument and can be omitted without undue loss by those who are not familiar with it.

2. THE INDUSTRIAL SETTING

Much of the book is concerned with industrial management. The effective management of technology is increasingly the key to company and national success. Of course, it is not the only ingredient of success, but it needs to be recognized and blended in.

Despite educational efforts to the contrary, one of the features of advanced technology has been the growing specialization of technologists. There is so much to absorb and learn in each field of technology that specialization is a natural route. However, in the long run, specialization narrows the breadth of vision and leads to parochialism. If there is too much of this, then perspective is lost and the perception of opportunities diminished. One of the objectives of this book is to restore the overview; that technologists and managers can better see their roles and contribution within a total framework.

The first group of readers envisaged for this book are the industrial managers, particularly in the technical and manufacturing functions. They may already be familiar with much of the subject matter, but their experience often comes in an unstructured, problem-driven form. In some ways, technology is akin to plants. Without a structure, their variety and nature is bewildering. But once you have a structure, a taxonomy, then what is before you falls into place. You can distinguish one from the other in a consistent manner. It is argued that a similar approach to technology is useful.

Industrial management here also includes those managers with non-technical roles, such as accountants, marketing and personnel staff. Their activities are often interwoven with technological problems and it will help them to appreciate their implications. Technology is not something to keep away from; it has form, structure and alogic.

The second group of readers are the aspiring managers: the technologists in post, the undergraduate and graduate students on technology courses or business programmes. For the preparation of their career they have already a good supply of management literature and teaching. However, the focus on technology is not so strong as, say, on marketing, finance or organizational behaviour. It is hoped that this book will widen the spectrum of such studies.

3. THE OVERALL CONTEXT

The third group of readers which this book hopes to serve is the informed and interested public. They may well be aware that although technology has furnished a standard of living never equalled before, most take technology for granted. It permeates our life at every level, yet we know little about its structure, width and sophistication. The fears about technology have had a better press.

It is not enough to talk about technology as if it were a 'lump'. In the 18th and 19th centuries one of the hallmarks of an educated gentleman was a classical education. It might be appropriate that one hallmark for an educated citizen in the 21st century is a fuller understanding of science and technology.

There is some concern about the limited public perception, interest and understanding of science and technology. For instance, in the United Kingdom, this has caused The Royal Society, The Royal Institution and The British Association for the Advancement of Science to join in 1985 in the establishment of COPUS, the Committee on the Public Understanding of Science. Its purpose is to encourage and promote the public understanding of science and technology and their impact on society. As far as technology is concerned, this book is a small endeavour in that cause.

Technology and technological change affect all sectors and activities of an economy, be it manufacturing, services, energy, or education. Many regard technology as the engine of growth. The economic power of Japan and the rise of new industrial nations in East Asia owe much to its consistent application and development. Competitive technology determines international trade and manufacturing employment.

This is too important a subject to be overlooked at the public policy level. Technology impinges on many policy decisions, yet if its substance, implications and second-order effects are not fully appreciated the risks of unexpected and, at times, undesirable consequences are substantially increased. Value judgements are involved and these are the stuff of politics. When they are made about the use and the future direction of technology this ceases to be the private domain for engineers and scientists.

4. THE NATURE OF TECHNOLOGY

On the basis that it is difficult to manage or develop technology without some knowledge of its structure and contents, Part One of the book is concerned with the concept, nature and constituents of technology. It sets it apart from science and industrial crafts. Its dynamic nature is the subject of technological change in Chapter 2.

As in a commercial context economic expectations are the main driver of technology development, the economics of technology warrant more attention. Chapter 3 introduces the subject of technological economics. It also outlines concepts of economics which are important to the technologist.

5. THE MANAGEMENT OF TECHNOLOGY

Part Two is the substance of the book. It stresses the importance and relevance of technology as a strand of corporate strategy. Both the development of product technology and the adoption of new manufacturing technologies are an expression of such a strategy. Its implementation is mainly in project form and Chapter 8 stresses some relevant aspects of development project management. The investment in, and the accounting for, technology are associated subjects and these will be discussed in Chapters 9 and 10.

Technology can be regarded as a commodity and with the escalating costs and risks of research and development its transfer is becoming increasingly important. The nature of this process is the 'gist' of Chapter 12. Allied to this is the question of 'appropriate technology' discussed in Chapter 11.

The management of improvement is closely linked to the intelligent use of technology and the opportunities of incremental change. Chapter 13 discusses the culture and the learning processes required. An aspect of technology management, given only modest attention so far is the role of standards and standardization, developed in Chapter 14. With the rapid development of technology in such fields as electronics and communication systems, standards have become increasingly important. Chapter 15 deals with the final theme of Part Two; the role of technology as an instrument of competition. This has become a major challenge not only to companies but economies.

6. TECHNOLOGY AND GOVERNMENT

In many ways the success of a national economy is the sum of individual and company achievements. Any government which wishes to sustain and enhance the welfare and living standards of its people has to formulate policies to this end. This also applies to the use and advancement of technology. Chapter 16 introduces this complex and, at times,contentious subject of public policy.

Paul Lowe
Edgware, Middlesex
January 1995

Acknowledgements

This book has drawn its material from a wide range of sources. In nearly all cases the references are given in the text, but I wish to add thanks to formal acknowledgement. In particular, I wish to thank the *Financial Times* and *The Times* for allowing me to quote from their articles, Rolls-Royce plc for providing information about their aeroengines, Hurco (Europe) Ltd for details about their machining centres and the British Standards Institution for their information about the development and updating of British Standards.

I also wish to express my gratitude to colleagues in the Department of Manufacturing and Engineering Systems at Brunel University, especially Professor M. Sahardi and Mrs Edgecock for their support when I was struggling. Furthermore, many thanks are due to Dr R.J. Grieve and Mr S. Sharma for their advice on how to handle a temperamental computer and on the finer points of word processing and computer graphics.

Last, but not least, I am grateful to my wife for her patience while this book was written.

PART ONE

The Nature of Technology

CHAPTER 1

The concept of technology

1.1 INTRODUCTION

The management of technology requires knowledge of technology. This means at least some awareness of what technology is about and where it fits within the totality of knowledge. This chapter is concerned with the nature, scope and structure of technology. It will enable the reader to judge what is technology and what is not. The details of a specific technology are left to the practitioners concerned. The objective here is to describe an overall framework and perspective for the evaluation and classification of technology. The difficulties of definition and the drawing of boundaries in a fuzzy and changing subject area are acknowledged.

1.2 THE NATURE OF KNOWLEDGE

It is the human experience that knowledge accumulates and advances. Not always at a regular pace – sometimes knowledge may also be lost – but within recorded history this advance has been continuous. We are concerned here particularly with a kind of knowledge that is general and regarded of value by mankind: knowledge of science and technology.

The immensity of this knowledge can only be utilized effectively if it can be expressed in terms of concepts and structure. A format, a system of classification is needed. The never-ending growth of such knowledge is a challenge to the structure which is based on knowledge previously acquired. New knowledge may supplement, extend or supplant previous knowledge and the framework based on it. The advance of knowledge is not only an addition of new facts but can entail changes and new relationships between its branches. In such cases, classifications, such as the Aristotelian system of science, no longer provide an adequate view of the perceived reality and fall. New classifications emerge only to be superseded in turn as knowledge expands.

Vickery (1975) describes the key features of a developed classification in the following manner:

1. The total universe of entities may be divided into broad **fields**. Physics, for instance, can be divided into heat, light, sound and electricity.
2. Each field can, in turn, be divided into **facets**. For example, the field of mechanical engineering, known as hydraulics, can thus be divided into hydrostatics, the flow of liquids, notches and weirs, pumps etc.
3. Each facet can be structured into an **hierarchy**, subdivided stage by stage by the application of a series of **characteristics**. These are then applied in an ordered sequence such as: engines, petrol engines, air-cooled petrol engines etc. Any one level of subdivision gives rise to a group of terms that constitute an **array**; for instance, petrol engines, steam engines, diesel engines etc.
4. Rules can then be applied to combine terms from the same array, from different arrays in the same facet, from different facets in the same field and from different fields. The rules usually involve the use of special relational operators or role indicators.
5. Each field, facet or term may be coded to fix its position in the whole system and to facilitate unambiguous combination with other codes.

In essence, we have here the framework of a taxonomy. These principles of classification provide structure, order and logical relationships for entities of knowledge. Taxonomy enables us to put order into existing knowledge and provides a catchment framework for the location of new knowledge. A problem of classification arises when the new knowledge suggests new relationships which cannot readily be accommodated in the existing structure.

A developed classification is, of course, a primary requirement for librarians, the custodians of recorded knowledge, and information scientists. Without classification the maintenance and retrieval of established knowledge would be in jeopardy. To the librarian, classification is a basic tool of his profession and provides him with a structure and horizon.

Apart from these professional specialists relatively few have this awareness of structure and horizon. The immense growth of technology over the last, say, 200 years has resulted in a loss of perspective. A technologist working in a specific field is only familiar with his and relevant subject matter. A non-technologist, even with some science education, has little feel for the ramifications and interrelationships of technology; it remains basically unknown – a challenge.

1.3 ASPECTS OF CLASSIFICATION

Although the basic role of classification is accepted, the actual task of classifying, within the context of advancing knowledge, has its problems.

Knowledge has value because of its usefulness, which is primarily seen in terms of application. The assessment and location of a specific parcel of knowledge reflects scientific and educational consensus; i.e. those who use it and who teach it come to a conclusion where it fits in best within the overall scheme of knowledge. That parcel of knowledge then becomes a 'discipline'. It reflects a basic division of labour which facilitates teaching, professional development and practical application. The allocation of such a slot by these utilitarian criteria is not always a comfortable arrangement. Knowledge, by itself, is oblivious to utility and complications arise. Some subject matter could equally well feature in two or more disciplines. Just consider physics, chemistry, physical chemistry, biochemistry, biophysics, inorganic chemistry, organic chemistry, materials science etc. The property of materials could feature in all of these.

The first feature of classification is that disciplines are not watertight compartments of knowledge. They are heterogeneous and they overlap; their contents vary with time and place. For instance, the disciplines of physics and mechanical engineering both cover the subject of thermodynamics, yet their respective content may differ considerably. This is more than emphasis; in teaching and research one has to judge how to use limited time and resources to achieve particular objectives.

A discipline is not static, it can ferment, bring forth a new discipline. In the 1940s electronics was little more than a minor constituent of electrical engineering. In the 1990s it is a major discipline in its own right. The growth of knowledge in a particular field may also encourage specialization, such as within biology where there has been the development of microbiology and molecular biology. On the other hand, with time and new knowledge some existing knowledge may become redundant, i.e. it no longer has the value it previously possessed. Again, the growth of knowledge can be particularly strong at the edges of two well established disciplines and may lead to the fusion of a new discipline, such as biochemistry.

A discipline need not be related to the laws of nature; it can be, as Ranganathan (1955) puts it, the distillation of practical activity and experience, such as the subjects of management or research method-ology. Similarly, there can be a clustering of subjects into coherent fields, such as manufacturing technology, which deals with the manipulation of materials, such as metal cutting, metal forming, the extrusion of polymers etc.

Further subject development can take place by a process of lamination where a primary subject can be expounded by a particular system of thought, such as Marxian economics; or within special boundary conditions, e.g. tribology; or a particular environment, such as marine biology.

1.4 THE CONCEPT OF TECHNOLOGY

Technology is a constituent of the universe of knowledge and shares the same problems of classification. An arbitrary but convenient location of technology in the continuum of knowledge can be indicated as follows:

Science | Technology | Know-how | Industrial art | Crafts

Classification is affected by problems of definition. Where does technology begin or end? Technological knowledge can be highly scientific and abstract; it can also be very concrete and empirical. Are there boundaries and if there are, do they stay in place?

Science is public and universal. The borderline between science and technology is fuzzy and dynamic. Technology is both public and private. The private component is marketable. Like industrial art, it can have a strong personal content.

Furthermore, the purpose of technology differs from that of science. The objective of science is to obtain general and publishable knowledge. In economic terms its findings become a public good. Technology, on the other hand, focuses on opportunities, specific problems or groups of problems. The solution of such problems often yields proprietary knowledge that may have considerable value. The owners of such intellectual property will only share this asset on commercial terms.

1.5 THE MEANING OF TECHNOLOGY

It is commonplace to talk about technology; the term is so all-pervasive that it is often taken for granted. Some regard it as the means by which man extends his power over his surroundings. For the purpose of this book this is too general a concept.

The first thing to note about technology is that there is no agreed meaning of that word; none of the clarity of a mathematical term or a physical constant. We are driven to the common usage of language, to the dictionaries. The word 'technology' comes from the Greek 'tekhnologia' which means the systematic treatment of an art or craft. (techne – is an art or skill; logia – is science or study). Based on this *The Oxford English Dictionary* considers that: 'Technology is the scientific study of the practical or industrial arts'. This definition has the disadvantage of imprecision, which stems from its historical pattern of usage. It remains a limited and somewhat blunt tool.

The French and German equivalents: 'technologie' and 'Gewerbekunde' have a similar origin. The German term 'technik' describes an industrial art and includes its principles, rules, know-how, manipulation and experience.

It also follows from such a definition that an enquiry into the nature and characteristics of technology is a second-order study in so far as it is, itself, an analysis of a study. The first-order study, i.e. technology itself, is concerned with the industrial arts; it uses scientific method in areas which have not always yielded completely to scientific enquiry. The 'art' element reflected in craft know-how is still important in some industries.

The etymological references to technology in *The Oxford English Dictionary* are of the 17th, 18th and 19th centuries. Technology is referred to as a description of the arts, 'especially the mechanical'. Subsequent references widen the use of the term 'technology' but do not go beyond 1882. This lack of more recent references in an authoritative dictionary is significant. The yield of scientific progress since that year and its effects on the 'industrial arts' is, to say the least, inadequately reflected. The definition implies the prior existence of these arts which are then subject to scientific scrutiny. Many of the current technologies, such as atomic power generation, the manufacture of petrochemicals or computers are the derivatives, the projection of scientific developments: the 'industrial arts' followed scientific enquiry here, they did not precede them.

The historical emphasis on the industrial arts fails to give sufficient emphasis to technology in what have become the service industries. Consider the level of technology serving electronic data interchange, the instantaneous world-wide booking of flights or hotel rooms, the operation of real-time global financial markets.

A number of interpretations of the term 'technology' have become established. For instance, Cornwall (1977) considers that the technology of a country at any point in time is the stock of knowledge that pertains primarily to the production of goods and services. The operational part of this stock of knowledge consists of a set of techniques, each technique being defined as a set of actions and decision rules, for transforming inputs into outputs.

Rosenberg (1982) argues that technology is more than the mere application of prior scientific knowledge. It is a knowledge of techniques, methods and designs which work even if, at times, the reasons why they work cannot always be explained. Technological progress does not necessarily require a full understanding of the underlying scientific principles. Technological knowledge may be accumulated by trial and error and often precedes scientific understanding. For instance, gentlemen farmers in the 18th century developed the cross-breeding of sheep and cattle without the benefit of genetics and still prospered as a result. Of course, scientific activity may subsequently be devoted to the understanding and explanation of the principles underlying an established technological phenomenon. Another important contribution of technology, in its interaction with science, has been the development of artefacts, such as instruments, which were crucial to scientific advance. Where would astronomers be without telescopes?

Dosi (1984) defines technology as a set of pieces of knowledge, both practical and theoretical, know-how, methods, procedures and physical devices which incorporate such knowledge. Disembodied technology consists of particular expertise, based on past attempts and technological solutions. It is akin in many ways to, and, of course, includes computer software.

Dosi's definition prompts the question of quantum. How big a set of pieces make a technology? One pragmatic answer is suggested by Langridge *et al.* (1972) who interpret technology as a sufficient body of knowledge or industrial practice to provide a lecture course at a final year university degree course or at master's level.

Transcribing Kuhn's interpretation (1962) of a scientific paradigm, Dosi sees a **technological paradigm** as a 'model' and a 'pattern' of solutions for specific technological problems. This has been applied by an established body of practitioners and based on principles derived from the natural sciences. Furthermore, it can be argued that there is an 'intellectual culture' to a technology, which is expressed by a style of thinking and language. Similarly, a **technological trajectory** is the pattern of 'normal' problem-solving activity based on a technological paradigm. Dosi emphasizes the focus of attention which is generated by technological paradigms.

Some writers have confined the term 'technology' to equipment and apparatus while others, such as Woodward (1965), consider it to be the collection of plant, machines, tools and methods available at a given time for the execution of the production task. Her concept includes operations technology, the knowledge technology, the tasks required by operators and the control system to be used by management. Such an explanation accords to a technology a number of very different characteristics, the breadth of which make it difficult to use the term for analytical purposes. With the object of reducing this problem Winner (1977) considers a technology to have three separate properties: equipment, technique which is the skill and knowledge required to use it, and organization which refers to the structure of control and coordination it requires.

Again, Jantsch (1967) in his report on Technological Forecasting to the Organisation for Economic Cooperation and Development (OECD) defines technology in the broadest sense:

Technology denotes the broad area of purposive application of the contents of the physical, life and behavioural sciences. It comprises the entire notion of technics as well as the medical, agricultural, management and other fields with their total hardware and software contents.

1.6 THE CHARACTER OF A SPECIFIC TECHNOLOGY

Technology can be seen at two levels. The first is the totality of technology, as it generally appears in public discussion. The second is a specific technology. While general concepts have their value, for the purpose of analysis and management the latter is often more useful. However, this introduces boundary problems. How can you describe or classify a specific technology? Where does it start and where does it finish? What are its distinguishing features? Where are its boundaries with science or the industrial arts?

There are a number of attributes that can be considered as part of a particular technology, such as:

1. a body of scientific principles and laws,
2. a related group of applications,
3. a specific set of artefacts,
4. specialist knowledge expressed in a cluster of techniques for investigation, measurement and application.
5. operational experience and know-how,
6. organization, expressed in structure and systems.

The constituents of a specific technology are shown in Fig. 1.1. The listed attributes suggest a working definition of technology which will be used in this book.

A technology is the structured application of scientific principles and practical knowledge to physical entities and systems.

A technologist, by virtue of his education, training and experience, will be able to extend and deepen the applications of scientific principles in his field.

A technician will be able to apply scientific principles and practical knowledge in his field.

A craftsman will be able to apply practical knowledge in his field.

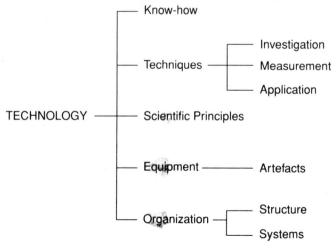

Fig 1.1 The constituents of a technology.

Attribute assessment allows the scrutiny of any parcel of knowledge to establish whether it constitutes a technology, an industrial art, or a craft. It also helps to relate a technology to other technologies. Most of the constituents are essential; certainly the scientific principles are, even if they are not always fully established in the early stages of a new technology. Again, with an emerging technology structured experience may not yet be fully developed.

One of the practical ways to depict a technology is to look at those who practise it. The practitioners are normally an identifiable group of technologists. They are often members of a particular professional organization and have completed a distinct course of education and training. Their affiliation reflects a division of labour and in an economic sense they have limited transferability, e.g. a metallurgist is unlikely to work as an electronic engineer.

A technology contains a body of formalized, transmittable scientific knowledge. An industrial art does not always have this and its acquisition is more of a practical task than an intellectual activity. Also, much of what is described as 'know-how' comes into this category. It is often specific to people but can substantially contribute to the successful application and growth of a technology.

It must be remembered that there is no unique technology. Particular scientific principles and techniques are often shared. Indeed, the very transfer of a technique from one technology to another often give rise to technological progress.

Technology has intellectual content and its use presumes a certain level of education and training. A technology can also contain intellectual property in the legal sense. It has utility and therefore value. That value depends on a context of application; technological change can alter it. This can occur within the spectrum of the particular technology or because of the development of other technologies.The value of the technology, and of those who practise it, is also a function of supply and demand.

1.6.1 The expression of technology

Technological knowledge is not as well articulated as scientific knowledge. Some of it is contained within skills and experience. This is what Polanyi (1958) calls 'tacit knowledge', i.e. knowledge that is acquired through practice and which cannot be stated explicitly. There is also the ownership of such knowledge; some of it may even be deliberately kept secret by specialists and industrial companies.

For instance, one major car company had a particular problem with the consistent evaluation of prototype component tests. Engineers carried out their assessments on the basis of their personal professional judgment. This 'knowledge-based' judgement led to inconsistencies between engineers and to training problems when staff left. In this case it was important to make the technology explicit, to make it 'rule-based', to put it into a manual format.

1.6.2 Disembodied technology

A distinction is often made between embodied and disembodied or 'soft' technology. Embodied technology is encapsulated in products and physical equipment, such as manufacturing plant. The technology has what Glasser (1982) calls, a technical root structure which gives a product or process its performance and application characteristics. Some writers include in embodiment the training or retraining of labour.

A disembodied technology is mainly intangible; there are no specific products which give it its particular character. The scientific principles underlying disembodied technologies are the social or management sciences rather than the natural sciences. Industrial engineering and quality assurance are typical examples of a disembodied technology. Its practitioners use artefacts but these are generic, such as an operating manual or a computer software package.

1.6.3 Generic technology

Another facet of technology is what Coombs *et al.* (1987) describe as its service characteristics. These, in essence, are the applications and uses of a technology. The greater the range of applications, the more generic is the technology. As distinct from industry or process specific technologies, generic technologies, by their very nature, have a wide range of application throughout industry. Lubrication, electronic device design, optoelectronics and robotics are typical generic technologies. Because of their wide scope of use, incremental improvements in such technologies can make significant contributions to technological progress in a wide range of industries.

1.6.4 Systems technologies

The rapid growth of information technology has in the last two decades yielded unprecedented scope for systems integration, such as with multimedia telecommunications. This is a fusion of hardware, such as computers and telecommunication equipment, software, such as digital systems and programming, and transmission networks, consisting of optical fibre links and interfaces. The constituent technologies have been described by Tassey (1992) as systems technologies. The feature of systems technologies is their great variety of possible combinations leading to a large number of different applications. Whereas a number of technologies embodied in a car are harnessed for a specific product range, with systems technologies it is the generic opportunities that make the impact.

1.6.5 Infratechnologies

There are groups of technologies which generally support the R&D, production and marketing of an industry. Expressions of these infratechnologies are measurement and test methods, calibration procedures incorporated in technical standards, published scientific and engineering data. They are important not only for the effective execution of R&D but also for the meaningful transmission of findings. In turn, they assist with the diffusion of new technologies.

1.7 THE SCOPE OF TECHNOLOGY

If it is difficult to define a particular technology, it is even more of a challenge to delineate the totality of technology. One way to appreciate the diversity and interlocking nature of technology is to look at the organization of the professional institutions concerned with various fields of technology, such as engineering, one of its major constituents. There are currently no less than 46 professional institutions associated with the UK Engineering Council. While there are elements of overlap in the professional practices of the different institutions they have, nevertheless, distinct fields of application. Also, many of the larger institutions have a number of specialist divisions, each concerned with a range of technological fields. For instance, some of the divisions of the Institution of Electrical Engineers are: computing and control; electronics; management and design; manufacturing; power.

The divisions, in turn, are composed of professional groups; e.g. the electronics division has groups concerned with measurement and instruments; electromagnetic compatibility; microelectronics and semiconductor devices; image processing and vision; signal processing; satellite communication. In all, there are 15 such groups in this division alone.

Furthermore, there are professional institutions whose activities are within both science and technology. For instance, the UK Institute of Food Science and Technology (IFST) is concerned with the applications of science and technology in improving the understanding of food in all its aspects. Its activities reflect the integration of a coherent body of knowledge, delineated as 'food science' and its applications as 'food technology'. Food science includes such subjects as biochemistry, microbiology and environmental health. The spectrum of 'food technology' continues to evolve and incorporates aspects of the total food production, distribution and consumption process. The IFST mission is governed by the role of food in modern society and will change as that role develops. IFST exists because there is a need for it which is not satisfied by individual science or technology disciplines. Its education and training programme is an example of subject classification resulting from society needs, not from any inherent structure of knowledge.

1.8 EXAMPLES OF THE CLASSIFICATION OF TECHNOLOGY

1.8.1 The Universal Decimal Classification

The nature and scope of technology is revealed by the various classification systems. One good example of this is the British Standard BS1000M: Part1: 1985, Universal Decimal Classification (UDC). This standard has been published by the British Standards Institution (BSI) at the original joint request of the British Society for International Bibliography (BSIB) and the Association of Special Libraries and Information Bureaux (ASLIB) with the approval of the Lake Placid Education Foundation, New York, the proprietors of the Dewey Decimal Classification, from which UDC derives.

It will be seen from Table 1.1 that this classification system is essentially numerical and uses digit values and positions to structure hierarchies of information for all levels of detail.

Table 1.1 Example of the classification structure

Class	Content
6	Applied sciences: medicine, technology
62	Engineering sciences
621	Mechanical and electrical engineering
621.3	Electrical engineering
621.32	Electric lamps

1.8.2 The Standard Industrial Classification

Another way to look at the totality of technology is to consider the spectrum of related industries in an economy. As an example, the 1980 Standard Industrial Classification published by the UK Central Statistical Office (1979) illustrates the ramifications of industrial activities and arts. It will be noted from the following divisions in Table 1.2 that the term 'industrial activities' is broadly interpreted. The total range consists of single-digit divisions.

Table 1.2 Overall industrial structure

Digit	Division
0	Agriculture, forestry and fishing
1	Energy and water supply industries
2	Extraction of minerals and ores other than fuels, manufacture of metals, mineral products and chemicals
3	Metal goods, engineering and vehicle industries
4	Other manufacturing industries
5	Construction
6	Distribution, hotels and catering, repairs
7	Transport and communication
8	Banking, finance, insurance, business services and leasing
9	Other services

Each division is subdivided into classes (second digit), the classes into groups (third digit) and groups into activity headings (fourth digit). Altogether there are 10 divisions, 60 classes, 222 groups and 334 activity headings. In addition, suffix numbers are used for further subdivisions. An example of the detailed digit structure is given in Table 1.3. These classifications indicate the range and structure of equipment and activities which reflect embodied technologies.

Table 1.3 Detail of classification structure

Digits	Meaning
3	Metal goods, engineering and vehicle industries
32	Mechanical engineering
322	Metal-working machine tools and engineers' tools
3221	Metal-working machine tools
3221/2	Metal-forming machine tools such as forging and swaging machines, presses

1.8.3 The main patent classifications

Another relevant framework for the assessment of the scope of technology is provided by patent classifications. In essence, a patent is a temporary monopoly granted for the exploitation of an invention. The totality of patents does not equate to the total of technology, because not all inventions are patented. Nor does a patent, as such, necessarily indicate the degree and the importance of an invention. Also, as with patent publication there are the risks of disclosure and subsequent infringement. Some firms with specialist process know-how prefer their own methods of secrecy rather than advertise their technological progress to the world at large. Nevertheless, the sum total of patent applications and granted patents, whether current or lapsed, gives, for practical purposes, a good insight into the extent and structure of technology. The following are the more important classifications.

The International Patent Classification (IPC)

The sixth edition of this classification consists of eight main sections which encompass the whole spectrum of technology, the industrial arts and applied science. Its diversity is expressed by 118 classes, 620 subclasses and about 67 000 main and subgroups.

The United Kingdom Patent Classification

`This is a detailed classification divided into eight main sections which broadly correspond with those of the International Patent Classification.

The subject matter of each section is divided into a number of divisions. There are altogether 40 divisions, each with its hierarchical structure within which patents are grouped. For instance, Section B 'Performing Operations' contains Division B3 'Working Metals', which, in turn, includes Heading B3M 'Rolling Metals'.

The US Patent Classification

Administered by the US Patent and Trademark Office (PTO), this is a most comprehensive and detailed classification system and contains about 113 000 subject categories. To handle such a daunting amount of detail, the patent file information is presented in such a manner that nearly every significant development in almost all technical fields flows in a natural time series sequence. The objective of this classification is to increase the awareness and use of the US Patent File as a prime source of technological information. The patent file has altogether about 25 million disclosures of technical innovation and probably all the major and most of the minor advances in technology that have occurred in the last 200 years.

1.9 THE SCALE OF TECHNOLOGY INFORMATION

We have already noted that patent applications and grants are an approximate measure of the growth and scale of technology. The level of patent applications in the 1980s was about 300 000 per year world-wide and this yielded about one million patent documents covering an average of 8–10 pages each. The accumulated store of such technology information is the largest in the world. According to Höusser (1981) only about 5–15% of the new knowledge described in the patent documents, whether they relate to devices, methods or materials, is contained in other scientific and technical literature. This points to a vast store of relatively untapped knowledge. Höusser quotes a then West German study that among 27 options for acquiring technology, patent documentation only ranked 14th in importance.

Since this survey the major developments in information technology, particularly with databases, networks and access by compact disc read only memory (CD-ROM) stations have made investigations relatively simple and may well encourage a greater use of patent information in the future.

1.10 LEVELS OF TECHNOLOGY

In addition to the description of the scale and extent of technology it is also possible to describe its practice in hierarchical terms. A technology can be evaluated on a scale which starts from the practical and repetitive at one end to the complex and abstract at the other. For instance, Parker (1982) gives six levels of technology relevant to the development of a new product. He outlines an hierarchy in terms of the problem-solving task required, the staff needed for product development and the publication level for the problem solution. Level 1, for instance, is a repetitive solution involving simple choices and is generally carried out by craftsmen. At best there would be a reference in a trade journal. This is at the margin of a technology. Level 6, on the other hand, involves precisely formulated unambiguous highly technological goals; new knowledge, complex systems, sophisticated tools of measurement and analysis and a power of abstract thinking. These are the hallmarks of an 'advanced technology'. It requires engineers and scientists of international reputation and would yield sufficient papers in 'prestige' journals to justify a new textbook.

Table 1.4 illustrates a modified approach based on five levels. Pivotal staff are those on whom the effective use of the technology depends.

Table 1.4 Features of different levels of technology

Level of technology	Characteristics and context of technology	Pivotal staff
1	simple artefacts and techniques, repetitive activities, mainly craft know-how, rudimentary use of scientific principles	craftsmen
2	main technology embodied in equipment, some technical know-how and application of scientific principles	technicians
3	considerable process and product know-how, some technology development, use of established techniques	graduate technologist
4	extensive know-how, equipment with advanced technologies, substantial R&D programmes, use of advanced commercial techniques	technologist professional reputation
5	global technological leadership, fusion of advanced technologies, extension of science base, strategy and organization to extend competitive advantage	technologist international reputation

1.11 TECHNOLOGY PORTFOLIOS

Most products and manufacturing processes incorporate a number of technologies. Some are key technologies which, in essence, are the very hallmark of the product. Other are subsidiary and scarcely known to the layman. Some of the technologies have changed little over decades; others have radically altered over a relatively short period of time. Often they have different life cycles and are at different stages in that cycle. Consider, for instance, some of the technologies embodied in a passenger car:

- steel making
- metal forming
- assembly operations
- combustion engineering
- control engineering
- electronic engineering
- glass technology

- metal cutting
- welding
- manufacturing automation
- mechanical engineering
- electrical engineering
- rubber and plastics technology
- paint technology

This list of 14 technologies, which is not exhaustive, indicates the range of technologies involved; the term 'automobile engineering' does not adequately express the range of such a portfolio. The arrival of new technologies, such as electronic engineering, has added new product attributes, such as engine management systems, while others, such as plastic body components, have diminished the role of existing technologies involved in the manufacture and manipulation of sheet steel. Thus, the advent of a new technology or a change in a constituent technology can affect other technologies in the portfolio as well as the life cycle of the product in which they are incorporated.

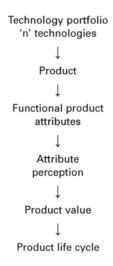

Technology portfolio
'n' technologies

↓

Product

↓

Functional product
attributes

↓

Attribute
perception

↓

Product value

↓

Product life cycle

Fig. 1.2 Product technologies and product value.

At any given period a key technology will be the driving technology and will govern the pace of development of the product as a whole. However, as a product or process evolves the driving technology can change. This is illustrated by the transformation of metal-cutting machine tools from about 1970 onwards. Before then the driving technology was the cutting process. After that it became the computer control of these machine tools.

The fusion of technologies within a product technology portfolio is shown in Fig. 1.2. The interdependence of technologies within a product portfolio can be important when it comes to technology choice. Unless the interrelatedness between apparently separate fields of technology has been assessed, product development can be affected by technological bottlenecks. The progress of one technology is then held up by the limitations of another. For instance, the development of solid oxide electric car fuel cells is currently affected by problems of product lifetime ceramics.

1.12 TECHNOLOGY AS AN ENVIRONMENT

Some writers, such as Amendola and Gaffard (1988), consider that technology need not just be seen as a specific way of, and equipment for, solving a particular problem. It can also be looked at as an environment characterized by specific resources that make it possible to devise and implement different solutions for different problems. The environment consists of public knowledge that underlies artefacts and the way in which they are used. In such a context the process of innovation consists of research and learning carried out in a given environment. This results in the emergence of entirely new skills and qualifications that bring about a modification of the environment itself. It thus makes it possible to extend the existing range of problems and solutions. Learning takes the form of the acquisition of new, and often higher, skills. Learning also takes the form of upgrading. The greater articulation of the human resource obtained in the carrying out of a particular process makes it possible in turn to define new production problems and new solutions to them.

An ultimate concept of technology is that of a socio-technological phenomenon, which goes much beyond equipment, labour skills and managerial systems. Within such a macro-view, technology includes cultural, social and psychological processes which are related to the central values of a country's culture. The strength of the managerial and social support systems is an important factor in the successful international transfer of technology.

1.13 SUMMARY

Technology is a branch of human knowledge which applies scientific principles and practical knowledge to physical entities and systems. It has an overall structure and classification. A particular technology contains a body of formalized, transmittable scientific knowledge and has distinct attributes such as specialist techniques for investigation, measurement and application. Its practitioners are often identifiable by education, training and professional affiliation. Most products and processes incorporate a number of technologies and can conveniently be evaluated in technology portfolio terms.

FURTHER READING

Langrish, J. *et al.* (1972) *Wealth from Knowledge*, The Macmillan Press Ltd, London.

Rosenberg, N. (1982) *Inside the Black Box*, Cambridge University Press, Cambridge

Vickery, B.C. (1975) *Classification and Indexing in Science*, 3rd ed, Butterworth, London.

CHAPTER 2

The nature of technological change

2.1 INTRODUCTION

Chapter 1 has been concerned with the description, structure and scope of technology as a whole. One of the problems encountered was its ever-changing subject matter. These dynamics of technology are the subject of this chapter. They are viewed in the context of business management, where technological change has to be justified by economic argument. It shows how invention and innovation contribute to technological change and the emergence of new technologies. It introduces the concept and the key stages of a technology life cycle. Technological progress is reviewed in terms of key parameters and their trajectories. The acceleration of change has produced technological discontinuities which challenge established businesses.

The diffusion of a technology is examined and the factors which govern its spread are indicated. The role of bibliometric patent analysis is put forward to forecast the diffusion between, and the combination of, technologies. Much of current technological progress is seen to be the result of technology convergence.

2.2 THE MEANING OF TECHNOLOGICAL CHANGE

It is helpful, in the first instance, to explain some key terms used in this and subsequent chapters. **Technological change** occurs where there is a change of, or an addition to, the underlying scientific principles which give a specific technology its particular character. **Technical change** is generally confined to changes within one or more of the other constituents of a technology, particularly to techniques and know-how. Technological change can produce technical change; the inverse is less likely. Although the latter is a subset of the former, the cumulative effect of technical change within its technology can over time transform the application and the value of a technology.

Technological progress describes the increased capability of a new or existing technology to satisfy human wants for goods and services. In commercial terms, it provides increased customer value. Technological progress comes in two basic ways. The first is the further development of existing product and process technologies. This can be anything from a small refinement in product design and the working practices in well established or 'mature' technologies to substantial changes where the technology is still relatively new and its use has not yet been optimized. Either way, as Coombs, Saviotti and Walsh (1992) suggest, a technology progresses by the solving of a sequence of puzzles, which sometimes is guided by theory but often is achieved entirely empirically.

The second form is the emergence of new technologies. It is quite common for a new technology to impinge on existing technologies, such as the impact of microelectronics on electromechanical control systems. With many products and processes the advent of a new technology will lead to a **technology portfolio change**.

The main features of technological and technical changes are novelty and application. Generally, these comprise discovery, invention and innovation. The term 'invention' has a meaning in law, but for our purpose it is also taken to include all ideas, the applications of which improve industrial operations and product attributes. For instance, low-cost automation devices for production or quality assurance, which express much of the know-how of a manufacturing company are also contained in this concept. The cumulative effects of such improvements are, of course, important.

2.3 THE CONCEPT OF INVENTION

In a general sense, an invention can be described as the creation of a new device, process or technique. Its nature is defined in law by the UK Patents Act 1977. Section 1 (1) of the Act specifies that a patent may be granted only for an invention in respect of which the following conditions are satisfied, that is to say:

- the invention is new,
- it involves an inventive step,
- it is capable of industrial application.

It should be noted from Section 1(2) that a computer programme and the presentation of information are not inventions for the purposes of the Act. Section 2(1) indicates that an invention shall be taken to be new if it does not form part of the state of the art, while Section 3 specifies the meaning of an inventive step:

An invention shall be taken to involve an inventive step if it is not obvious to a person skilled in the art ...

It is interesting that the word 'technology' does not feature in the legal definition of the term 'invention'.

2.3.1 Patent profiles

Published patent statistics furnish trends in the pattern of inventions and are an indicator of technological change to come. An example of useful information here is the Annual Report and Accounts of the UK Patent Office. For instance, Appendix 4 of the 1991–2 report gives the numbers of patent applications and grants in the major sections of patent classification. Significant trends of inventions in the following fields are listed:

1. Silicon polymers
2. Development of anti-theft devices (social changes)
3. Biodegradable laminates (environmental pressures)
4. Single-use syringes to prevent cross-infections (problem of AIDS)
5. Refurbishment of existing turbine blades (extension of equipment life; cost reduction)
6. Control of radial clearance between tips of turbine blades and engine shroud to maintain a minimum clearance despite thermal expansion (improvement of operating efficiencies of gas turbine engines).

Trends 1, 5 and 6 are improvement in materials, design and manufacture which reflect longer-term R&D strategy. Trends 2, 3 and 4 show the response to opportunities provided by a changing world. The changes themselves are affected by value judgements and economic choices. Trend 4 is the response to strong public opinion, anxious to see a new problem resolved.

The trends in patents during 1992 have been reflected in new or improved products in 1993–4. The Annual Report of the Patent Office for 1993–4, highlighted, in turn, significant trends in the replacement of chlorofluorocarbons (CFCs), biodegradable polymers, low-emission vehicles, cellular and cordless telephones. It is interesting that the first three trends reflect public and government concern about environmental problems.

Again, reports published by the US Patent and Trademark Office, as part of its technology assessment and forecast programme, have:

1. pinpointed technological areas experiencing the most significant growth of foreign activity,
2. reviewed the patenting in high-interest technologies,
3. profiled across all technology the patent activity of selected countries and the largest corporate patentees,
4. presented substantive reviews of active technologies,
5. examined the international balance of patenting,
6. compared patent activity against various economic parameters.

2.4 THE NATURE OF INNOVATION

Innovation is the application of inventions to industrial or commercial purposes, typically to products and processes. It is their translation into manufactured and marketable products. Similarly, innovation is applied to the development and improvement of services. The context is economic – as one industrialist said tersely and optimistically: 'innovation is profitable change'. The timescale for the innovation process can be substantial and an innovation is only realized with the first commercial transaction involving a new product, process or system.

Innovation can also be in the form of a new arrangement of existing products and technologies. For instance, Dussauge, Hart and Ramanantson, (1992) suggest that a car can have dozens of component technologies. Innovations can occur with a particular component, as well as with the component combination within in a larger system, framework or 'architecture'. Innovation within a component technology is modular, if it does not affect the overall product structure. An architectural innovation is a change of product which has no significant effect on component technologies.

Just as technology has been interpreted in both specific and in broad terms by different writers so it is with innovation. Innovation can also include the organizational change and infrastructure needed for the large-scale use of artefacts. For example, cars require petrol stations, traffic control and road maintenance.

These various aspects of innovation are well encapsulated by the following definition in the Organization for Economic Cooperation and Development (OECD) 'Frascati Manual' (1993).

> Scientific and technological innovation may be considered as the transformation of an idea into a new or improved saleable product or operational process in industry and commerce or into a new approach to a social service. It thus consists of all those scientific, technological, commercial and financial steps necessary for the successful development and marketing of new or improved manufactured products, the commercial use of new or improved processes and equipment or the introduction of a new approach to a social service.

2.5 THE EMERGENCE OF NEW TECHNOLOGIES

New technologies are the result of a complex interaction between scientific advances, institutional factors and economic mechanisms. The first provides the scenario of opportunities while the others may be regarded as focusing and filtering devices.

It has been suggested that new technologies, as distinct from scientific discoveries, emerge at certain stages of 'long waves' of economic development. Waves of innovative activity are associated with long-term cycles of economic upswing and recession, known as the Kondratiev waves after the Russian economist who first observed this relationship. The economic upturn within each cycle is linked to the development and diffusion of new technologies embodied in new products. Historically, such waves have lasted 40–60 years. The current cycle, known as the fifth Kondratiev long wave, is presumed to have begun in the early 1980s, mainly as a result of the further development of computers, data banks, networks, telecommunications and satellites, optical fibres, manufacturing automation, biotechnology and advanced materials.

According to the Technology Administration Division, US Department of Commerce (1990), an emerging technology is one in which research has progressed far enough to indicate a high probability of technical success for new products and applications that might have substantial markets

Table 2.1 The emerging technologies

Emerging technology	Some major technology elements
Advanced materials	Structural and functional ceramics, ceramic and metal matrix composites, intermetallic and lightweight alloys, advanced polymers, diamond thin films, biomaterials
Superconductors	High and low temperature ceramic conductors
Advanced semiconductor devices	Compound semiconductors, memory chips, X-ray lithography
Digital imaging technology	High definition systems, data compression, image processing
High-density data storage	Magnetic storage, magneto-optical storage
High-performance computing	Modular software, numerical simulation, neural networks
Optoelectronics	Optical fibres, optical computing, solid-state lasers, optical sensors
Artificial intelligence	Intelligent machines, expert systems
Flexible computer-integrated manufacturing	Computer-integrated manufacture, Computer-aided design
Sensor technology	Active/passive sensors, feedback and process control
Biotechnology	Bioprocessing, genetic engineering, bioelectronics
Medical devices and diagnostics	Cellular-level sensors, medical imaging, fibre optic probes

(Source: Technology Administration Division, US Department of Commerce, 1990)

within approximately ten years. It must have the potential either to create new products and industries or provide large advances in productivity or in the quality of products manufactured by existing industries which already supply large, important markets.

Some emerging technologies, often encapsulated in new products, such as drugs, have a focused impact. Others are generic in that they advance the technical infrastructure or enhance the effectiveness of manufacturing processes. A 1990 survey by the Technology Administration Division indicates the most important emerging technologies. These are summarized in Table 2.1.

2.6 THE LIFE CYCLE OF A TECHNOLOGY

The biological concept of a life cycle can also be applied to a technology and the products associated with it. There is the equivalent of birth, growth, maturity and death.

When a new technology emerges, particularly as a result of a group of key inventions, the products and processes involved are not yet fully developed. The nature and the scope of the market are still tentative. The innovation context is fluid and there is a period of vigorous progress when innovations follow each other rapidly and new equipment evolves. A number of alternative product and process designs compete until a dominant design is established. A diversity of applications foster market growth and structure. This is followed by a period of consolidation when the emphasis is on standardization, manufacturing efficiencies and, where feasible, the economies of scale of mass production. The innovation context becomes specific. With time the technology and its products become 'mature'; the rate of innovation slows while further development is affected by diminished returns and rapid increases in costs. Markets become increasingly price-competitive and growth begins to level out. A natural limit is eventually reached when the cost of the further development of a product or process becomes prohibitive. Ultimately, the existing technology and products are replaced by a new technology and become obsolete. The main characteristics of such a life cycle are illustrated in Fig. 2.1. Ryan (1984) puts these stages into a convenient pattern of integration with product life cycles as shown in Table 2.2.

Dosi (1984) interprets the natural limit of a technology as a 'technological frontier'. This is translated as a set of the highest (or lowest) values of technological parameters achieved in a given economic context. For instance, in commercial aircraft a maximum value could be airspeed and a minimum value fuel consumption. Emphasis on the improvement of a particular parameter often depends on external factors, such as changing fuel prices or pollution legislation.

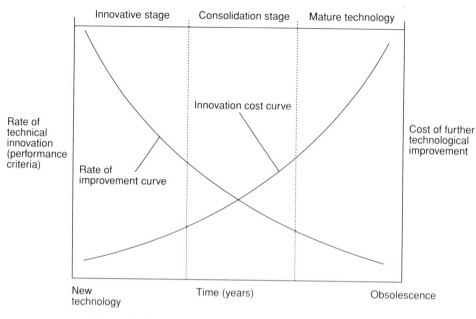

Fig. 2.1 The life cycle characteristics of a technology.

Table 2.2 Technology and product life cycle stages

Technological stage	*Feature*
Technology development	Basic technology
Technology application	Technology + application
Application launch	Technology + application + product launch
Application growth	Technology + application + product sales
Technological maturity	Technology + application + fall in product sales
Degraded technology	Minimal product sales + loss of application + alternative technology

(Source: Ryan, 1984)

Considerable attention has been given to the prospects of success of a new technology; whether the products it has created are more likely to succeed as a result of a 'market pull' or a 'technology push'. Striking examples of economic success and failure can be found with both categories. Where a technology is entirely new, such as with antibiotics in the 1940s, the technology generates its own demand. But later in the technology cycle when the technology is mature and competition is strong then the market pull may become more decisive to the success of the technology and its associated products.

2.7 THE MOTIVATION FOR TECHNOLOGICAL CHANGE

There are essentially two categories of motivation for technological change. The first is from within a business; the second is a response to pressures from its environment.

The chief characteristic of the first category is the perception of a business opportunity and the initiative from within – the hallmark of the entrepreneur. This is akin to Schumpeter's (1961) 'new combination of the means of production'. The opportunity is seen in terms of economic benefit in the form of increased, or maintained, revenues and profits. These are a function of: added value, and/or the reduced cost of providing goods or services.

Technological change can also be seen as a response mechanism. A business survives and prospers, just like a biological organism, by responding to the nature and the changes of its environment. These can encourage, or even compel, a firm to introduce technological changes. The most important sources of pressure are legal, economic and social.

For instance, changes in legislation, such as health and safety, or environment, will encourage, or demand technological change, such as with vehicle emission control. As far as a firm is concerned, technological change is then a response to a new context in which it operates.

Technological innovation through the development of new products, such as sophisticated electronic leisure equipment, is also fostered by a demand from higher levels of income. Similarly, social needs have stimulated the development of advanced security systems.

2.8 THE NATURE OF TECHNOLOGICAL PROGRESS

We have already indicated that technological progress relates to the increased capability of a new or existing technology to satisfy human wants for goods and services and thus to enhance their customer value. How is such progress assessed? For instance, for a product the criteria of assessment consist of one or more of the following:

1. lower costs for a given specification,
2. improved technological parameter values giving better functional performance, e.g. engine fuel consumption,
3. greater reliability,
4. increase in scale (load carrying capacity),
5. miniaturization (circuits on chip).

Basically, progress is achieved through discovery, invention and innovation. The scale, nature and impact of each of these three components can vary tremendously. For instance, an innovation can be anything from a minor, incremental improvement of an established manufacturing process to a breakthrough transformation of an industry. This is a prospect for electric power generation when fusion nuclear power is expected to replace the current atomic fission process.

An innovation can be regarded as 'major' when it changes materially the functions of a product or the basic way in which it is made. Abernathy (1978) suggests that such products mainly compete with their predecessors on their superior functional performance rather than lower initial costs; that is they incorporate performance-maximizing rather than cost-minimizing innovations.

2.8.1 The rate of innovation

The number of patents applied for in a subject area or the number actually granted in a specified period gives some indication of the rate of invention and as most patent applications are filed by corporations they are an output indicator of R&D activities. Because of subsequent development work the rate of innovation is seen as a lagged function of this.

Another approach is to take the sales value of products which are new in nature and new on the market. Typical macro-figures for the US economy in the 1960s gave an annual rate of innovation of 3–4%. There are however wide variations between industries, with the electronic and aerospace industries having innovation rates over 10%, while some companies in those industries exceed rates of 20%.

The rate of progress depends on the life cycle position of the technology and the resources devoted to technological problem solving. In turn, the applied resources are a function of the economic benefits expected and the contextual pressures of a given environment. The attitudes of a company to a perceived technological opportunity will depend on the competitive advantage it will yield.

2.8.2 Technological evolution

Although the rate of innovation may decline when a new technology has become established, progress often continues. According to Porter(1985) this may be due to two reasons. The first is industrial growth. A larger scale of operation may justify the further development of technologies. The second reason is learning curve effects. Extended experience brings out new ideas and develops know-how. There is the yield of greater familiarity with the technology in an industry, especially where there are established relationships between suppliers and users.

Many minor changes are mundane and are not linked with a technological breakthrough. They are therefore less obvious to competitors and can provide a company with a significant cumulative advantage. Such advantage can be sustained and broadened over industrial sectors and national economies as the Japanese have shown through continuous, long-term incremental process and product innovation.

The substantial benefits of technological evolution can be seen in electric power generation where over the last three decades there has been a large increase in thermal efficiency. While part of this is due to the economy of scale with larger power stations there have been continuous detailed improvements in boilers, turbines and generators.

2.8.3 Technological trajectories

Essentially, a trajectory is the path of progress of a technology embodied in a product or process. Glasser (1982) notes that it is governed by the critical problems of its root structure which limit its performance or further applications. The problems can be ranked in order of criticality and required development work. This results in a cluster of possible technological directions, bounded by the variables concerned. For instance, the application of computer-controlled systems to machine tools has been followed by computer-controlled test equipment. Similar skills, knowledge and experience were required.

Advances within a technological trajectory will eventually be affected by a combination of diminishing returns and increasing costs. As the technology becomes more and more specific the direction of the trajectory also becomes more uniquely defined. This will limit the range of options left available.

A well-known example of a technological trajectory is the modern jet engine. For instance, The *Financial Times* of 23 September 1987 listed a family of nine variants of the Rolls Royce RB-211 up to that date. The variants evolved from the original 22B engine and the significant technological parameter here is the engine thrust. Some of the RB-211 family have the following engine thrusts (lb):

RB211-22B	42 000 (1971, original engine)
RB211-524	48 000
RB211-524D4	53 000
RB211-524H	60 600 (1987)

By 1993 derivatives of the RB211-524L, the Trent 700 and 800 engines, achieved thrusts of up to 106 000 lb. This is an increase of 152% over 22 years for a key parameter. A report on this latest achievement stressed the roles of design (within the technology) and the use of advanced materials (fusion of technologies). Fig. 2.2 shows the trajectory for this parameter.

2.8.4 The multi-parameter evaluation of a trajectory

In the above example the key technology parameter was engine thrust. In other cases, such as with a family car, there can be a number of

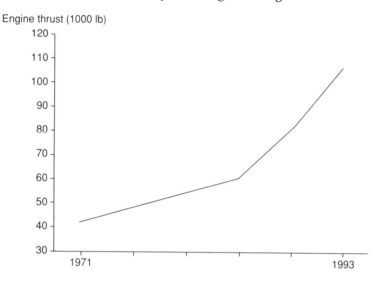

Fig. 2.2 Simplified trajectory of aeroengine thrust (source: Rolls-Royce plc).

parameters of importance. These could be competitive and time depen dent. For instance with a car, fuel efficiency can be just as important as speed or reliability. Girifalco (1991) suggests a 'figure of merit' (FOM) approach to the assessment of a technology trajectory with several signif-icant parameters. Its simplest form is as follows:

$$FOM = aA + bB + cC + \ldots$$

where A, B, C . . . are parameter values and
    a, b, c . . . are weightings given to each value.

To provide consistency between the parameters they are best expressed in a dimensionless or index form. Where the effect of technological progress results in the reduction of parameter values, such as for exhaust emis-sions, such parameters need to be expressed in an inverse form.

2.8.5 The analysis of technological change

As technological change can affect both a product and how it is made, Abernathy (1978) regards the 'productive unit' as a suitable unit for analysis. He defines this as an integral production process that is located in one place under a common management to produce a particular product. This could be, for instance, a car body plant run by a large vehicle manufacturer.

Abernathy groups process innovations into four functions: improved process capability; new process organization; the integration of an existing process and improving the overall process as a system. Furthermore, in his study of the history of the American automobile industry, he puts forward two major categories of innovation patterns.

In the first case, at the fluid stage, product design is subject to radical change. Product characteristics are in flux, the emphasis of product innovation is on improved functional performance rather than cost reduction. Production systems are flexible but inefficient and even major innovations may only have a small, immediate economic impact. In the second case, the product is standardized, change is incremental, information about needed product features is relatively visible and the economic impact of any improvement is large and immediate. As an industry and a product mature there is a transition from the first to the second category. This process of consolidation can, of course, take place over an extended time-scale and temporary intermediate patterns may appear. Fig. 2.3 shows the pattern of transformation.

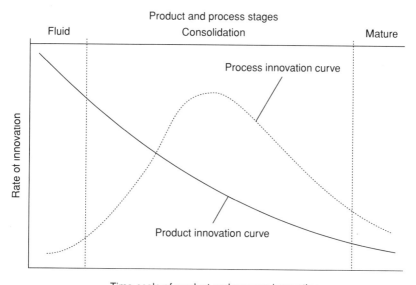

Fig. 2.3 The transformation of innovation stages.

2.8.6 A scale of technological change

One important aspect of the assessment of technological change is to determine how substantial it is. If a technology is established in terms of a body of knowledge and industrial practice the extent of changes within these will give some scale of the change. Langrish *et al.* (1972) used such a scale to analyse 84 innovations which obtained the Queen's Award to Industry in 1966 and 1967 for technological innovation. They assessed technological change in 'textbook' terms as shown in Table 2.3.

Table 2.3 A scale of technological change

Level	Extent of technological change
5	Innovation leads to a new technology; a new text book with a new title is needed.
4	Innovation makes several chapters of the standard book out of date.
3	Innovation requires major changes in one or two chapters or the addition of new chapters.
2	Innovation requires the alteration or addition of some paragraphs.
1	Innovation makes a very small difference to the text.

(Source: Langrish *et al.*, 1972)

Similar assessment could also be made about related product and process changes. As demonstrated by them, such a scale is usable, although there are limitations. It should also be noted that the extent of change is not necessarily linked to the economic worth of the innovations.

2.9.7   Technological discontinuities

This is a special case when one technology is rapidly displaced by another. Such displacements can have a critical impact on existing products and/or processes. Within a relatively short period there are rapid technical changes which are reflected by a substantial improvement of a key parameter, such as the speed of an aircraft, the efficiency of power generation or the rate of output of a manufacturing unit. In such fields as

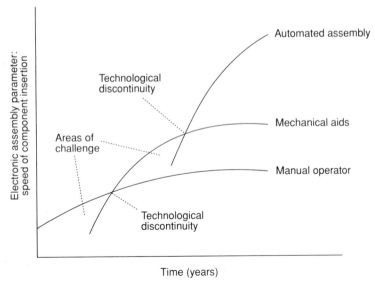

Fig 2.4 Technological discontinuities in electronic assembly.

electronic assembly, existing manufacturing plant rapidly becomes obsolescent and a company has to face substantial investment and skills acquisition programmes to stay competitive. Although it may not yet have the measure of the new technology and all its implications, it will have to adopt it or lose the business. Furthermore, where a new portfolio of competence is required it gives an opportunity for new firms to establish themselves with the latest technology. The risks associated with such periods of rapid transition generate a turbulent business environment. Figure 2.4 illustrates the nature of technological discontinuities for electronic assembly work.

2.9 THE NATURE OF A MATURE TECHNOLOGY

We have already seen that there is an ultimate limit to the growth of a particular technology. A ceiling or boundary is reached when both the diminishing returns and utility of further development work no longer justify its costs. This is reached when a key performance parameter, i.e. one that yields a specific customer value, becomes prohibitively expensive to develop further. The technology becomes R&D-inelastic with respect to key performance criteria. The rate of further innovation tails off.

The boundary can be a physical as well as an economic barrier. Because of the increase of turbulence near the speed of sound (Mach1) the transformation from subsonic to supersonic flight calls for non-linear increases in engine power. Again, the further development of one technology may be hostage to the barriers encountered with a supporting technology, such as the strength of materials at high temperatures.

The limits of a technology can also be linked to product maturity. In turn, this is related to market preferences. If a product matches market needs, then some regard it as mature.

There are interesting cases where the challenge of new technologies and products stimulates the defence of an incumbent, mature technology. Heavy investment in and past success with an existing technology tempts some firms to concentrate on its improvement, particularly where the challenge of the new technology is still tentative and the outcome is uncertain. The degree of support to be given to a challenged technology then becomes a major business problem.

For instance, Rothwell and Zegveld (1985) quote the case of the UK alkali industry when in the 1880s it concentrated on the challenged Leblanc process whereas European companies took up the new and more efficient Solvay process. This is akin to the sailing ship phenomenon where, because of the challenge of steam, there was more intensive development of sailing technology to combat the threat.

2.9.1 **Technological de-maturity**

Sometimes, products with a substantial portfolio of mature technologies acquire a new development impetus. This is due to the addition of a new technology which significantly improves product or process performance. Important examples of this are new polymer components for car bodies or automated robot welding lines for car body assembly. Also a new technology may yield novel product attributes for which there is a strong market demand, e.g. improved appearance, high reliability, or yield significant cost advantages. In some cases, such as with electronics, it gives the opportunity for a major product redesign and corresponding market development.

2.10 THE NATURE OF DIFFUSION

Diffusion is concerned with the spread of a new or improved technology and its related products. A new technology gives rise to innovations and these will spread to the industries where they find application. The diffusion may be industry specific or, if it is a generic technology, such as optical fibres, it may spread across a wide spectrum of industrial and scientific activities. Stoneman (1983) classifies the following levels of diffusion:

1. Intra-firm. This is particularly significant with transnational corporations.
2. Inter-firm. This applies to specific industrial sectors.
3. Economy-wide. This is typical of generic technologies.

The pattern of diffusion can often be expressed by an 'S' or logistic curve which shows in cumulative terms the percentage of companies in an industry that have adopted a new technology. This is well described by Nabseth and Gray (1974) who carried out detailed studies of the diffusion of numerically controlled machine tools, the float glass process and the continuous casting of steel. Mathematical models describing the rate of imitation of technological change have been formulated by Mansfield (1961) who investigated major innovations in the US iron and steel, brewing, railroad and coal industries during the period 1900–45. The development work for these industries was mainly carried out by equipment manufacturers and the innovations required substantial investments by the users. Of the 12 innovations studied, nine took over 20 years to diffuse completely in the industries concerned. This suggests that the path of diffusion for a new technology is not necessarily an open road. There are the capital costs of equipment which embody it and the initial operational problems that a user has to overcome. These can lead to an extended time-scale of diffusion in some industries where the previous technology is run out with the plant which incorporates it.

Mansfield considered that the nature of the curve is affected by the proportion of companies that have already adopted the innovation and compete with it against the 'hold outs', the profitability of the innovation and the investment required to adopt it. For instance, the greater the profitability of the innovation, the narrower will be the 'S' curve and *vice versa*. A typical 'S' curve is shown in Fig. 2.5.

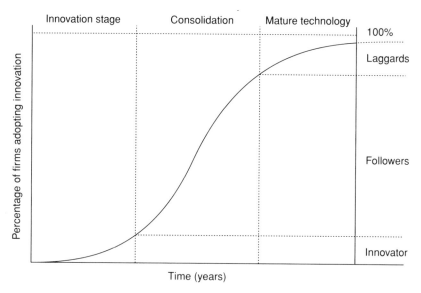

Fig. 2.5 The 'S' curve of diffusion.

Dosi (1984) stresses the need for care in the use of such curves. Firstly, there is the problem of the precise definition of the term 'innovation'. An innovation is not necessarily a once-for-all occurrence; there may be a sequence of significant incremental improvements. These are often the result of user feedback, such as application opportunities, as well as competition between suppliers. Diffusion is not a one way traffic; the innovator can also learn from the imitator. The innovation evolves as it diffuses. Secondly, there are the effects of price changes for the innovation itself. Thirdly there is the process of diffusion in the production, as distinct from the adoption, of the innovation. Last, but by no means least, are the capabilities and skill levels of the potential adopters. They need to recognize, implement and maintain the new technology.

There are various forms of diffusion. Cornwall (1977) classifies these as follows:

1. the diffusion of a product or a process,
2. the use and the production of a product,
3. the diffusion of technological knowledge as such.

The stock of technological knowledge can increase by:

1. learning from the production process itself;
2. developments in the various sub-branches of science, both pure and applied, where the basic knowledge is developed.

The rate of diffusion can be measured by the time interval between the use or production by the innovator and the initial adoption by an imitator. Useful time indicators are when 50% and 90% of all potential users have taken up the innovation. The nature and speed of diffusion is also affected by the source of the innovation. If, for instance, a process user generates a process innovation he naturally wishes to retain this to maintain and to enhance his competitive advantage. He has more control in limiting access to his process. If, on the other hand, an innovation is embodied in a new product its manufacturer will use it to enhance his market position and therefore spread the innovation as quickly as he can. The speed and extent of the diffusion of a technology, such as robotics, is further affected by the capability of prospective users to adopt it. There is also the cluster effect of related technologies which influence respective diffusions.

2.10.1 The concept of entrained innovation

Where there are interlinked stages of manufacture, irrespective of the ownership of the manufacturing process, one innovation within the process chain can cause an economic imbalance giving rise to bottlenecks. Because of the expected benefits which their removal may yield, attention will focus on these bottlenecks, giving rise to further innovations.

The classical illustration of this is the development of the UK cotton industry during the period 1760–1830, the Industrial Revolution. The inventions by Hargreaves in 1764, Arkwright in 1769 and Crompton, 1774 and 1779, led to the rapid diffusion of the new spinning machines. Their much greater speed and efficiency brought about an imbalance in the cotton industry. The carding of cotton, the main process prior to spinning, became a major bottleneck. It stimulated efforts to speed up the process and this was achieved by the technique of continuous carding invented by Arkwright and Hargreaves by 1785.

A recent example of entrained innovation has been the rapid development of semiconductor devices. This put pressure on the development of equipment and materials because of the need to fabricate more advanced devices. The corresponding development of high accuracy automated equipment is an example of entrained innovation.

2.10.2 Diffusion between technologies

One feature of technological progress is the transfer of knowledge from one field of technology to another. Knowledge diffusion can therefore be seen as the application of field-specific knowledge to other fields, as a function of time. Engelsman and van Raan (1991) developed the concept of 'knowledge-supplying' and 'knowledge-absorbing' fields of technology.

In their study of the interaction of technologies they encounter the task of defining and measuring the nature and degree of interaction between different technologies. The interaction between a field of science and its related technologies is, broadly speaking, the dialogue between expert practitioners in the same area of knowledge. They often share the same framework of education and training and interact in their professional activities. This is not necessarily the case where entirely different disciplines are concerned. Yet much of current technological advance is at the boundary of established disciplines.

Engelsman and van Raan use bibliometric techniques, mainly a citation analysis of patents and patent applications, to establish indicators and to assess the interactions between different technologies. Bibliometrics is that branch of information theory that attempts to analyse quantitatively the properties and behaviour of recorded knowledge. These techniques are used to support, not to supplant, the expert knowledge of practitioners in a given field. Where the patents reflect the spread of generic technologies, such as the application of carbon fibres, the diffusion process becomes particularly interesting because it spreads in an application area may not yet have reached a commercial level but can be predicted from the patent activity.

They distinguish three main types of science and technology indicators. The first is concerned with the size and other characteristics of science and technology outputs. The second illustrates their corresponding impacts; while the third refers to the structural features of science and technology. Two types of techniques are applied. One-dimensional techniques are primarily counts of specific bibliometric elements, such as key words in patents or citations. Two-dimensional techniques focus on the linkages between different technologies; the count is now of 'co-occurrences' of specific information elements, for example, terms appertaining to laser technology, found in patents in the fields of metrology and sensors.

They view technologies as complicated, heterogeneous conglomerates of different fields of activity, characterized by many interrelated aspects. To express these vast quantities of information in a numerical or tabular form entails the loss of perspective; there are too many details for the key aspects to emerge. Their overall approach is to map these relationships. A cartography of technology formats the data in a graphical form and reduces it to a more manageable level. A usable overview is facilitated and structures   can be evaluated with a practical level of accuracy.

Bibliometric methods are flexible and can be applied to different levels of aggregation. Where they are applied to the analysis of patents they are subject to the limitations of patent classification systems. These systems are for the use and benefit of professional practitioners; that they provide opportunities for research is fortuitous. It is however encumbent on the research worker to be aware of the caveats concerning patent analysis. Patents are a proxy indicator. They do not describe the sum total of a technology and patent applications may also be rejected where in the view of the examiners they come too close to existing patents. Nor does the data they yield allow for national, cultural aspects. Nevertheless, techniques such as citation or co-citation analysis, if used with caution, indicate the 'direction of influence', i.e. the nature of knowledge flow from one technology to another. This is important for the assessment and the forecasting of technological change.

On the basis of the International Patent Classification (IPC) Engelsman and van Raan carried out a citation analysis on the interactions each year of 28 technologies. Corresponding maps give a snapshot of the strength and nature of the linkages. A time series of such maps crystallizes trends of technological development.

2.10.3 Generic technologies: uncertainty and complexity

Whereas many technologies have natural boundaries of application, with generic technologies it is harder to predict their ultimate pattern of usage. Dodgson (1989) makes this point with electronics. It has a number of direct industrial branches, but at the same time it has an all-pervasive effect on other technologies, processes and systems. It is therefore relevant to consider the generic content of a technical development. Not only is the timing and impact uncertain but also the width of induced change.

This view is reinforced by Freeman and Perez (1988). For instance, the pervasiveness of information technology (IT) is the expression of a new 'techno-economic paradigm' where the changes involved go beyond the engineering trajectories for specific product or process technologies. They affect the input cost structure and conditions of production and distribution throughout the economic system.

2.11 TECHNOLOGY CONVERGENCE

Some innovations have to wait for other technologies to make them successful. However, rapid technological changes in separate fields can yield special strategic opportunities. With some advanced technologies a single technological breakthrough is insufficient; it is the fusion of several breakthroughs in different fields that may be needed. Kodama (1991) suggests that a fusion-type of research becomes possible with the joint efforts of the industries concerned. He cites the case of two particular development problems with optical fibres. The first concerned their high transmission losses. This was solved by a major electronics corporation by the use of longer wave lengths. The second problem was the mechanical fragility of the fibres. This was eliminated by the cable manufacturers who invented a new coating technology. Fusion-type innovation contributes not so much to the radical growth of a particular company but to the gradual growth of all the companies in the relevant industries. The essence of technological fusion is its reciprocity. However, the exploitation of a particular idea or invention also depends on the convergence of different technological capabilities in the firms which could develop suitable applications. Where their integration is a key requisite there is a growing emphasis on fusion, such as with mechatronics, which integrates mechanics and electronics.

A current convergence of technologies in the field of information technology is illustrated by the development of multimedia services and workstations. These provide speech, text, fixed and moving images across telephone lines. They will significantly change business operations by teleworking and video conferencing, while personal life styles will be affected by such opportunities as home shopping and banking. The key technologies involved are:

- computing and software engineering,
- digital techniques for the common base of information processing and transmission,
- opto-electronics for data transmission,
- radio engineering.

2.12 SUMMARY

New scientific principles lead to technological change while new techniques and know-how lead to technical change. New technologies alter the technology portfolio of products and processes. Invention is the creation of a new device or process and innovation applies these to products and processes.

A technology has a life cycle. Its path of progress can be shown as a trajectory embodied in products and processes. The initial stage is fluid, with many innovations and a range of designs. A dominant design or process evolves and consolidation leads to standardization, volume production and economies of scale. As the rate of innovation slows down, the technology becomes mature. The emergence of new technologies may render it obsolete. A technological discontinuity occurs where there is rapid replacement.

Diffusion describes the spread of a new technology and its related products. Diffusion between technologies is the transfer of knowledge from one technology to another and often leads to changes in product technology portfolios. Technology convergence integrates rapid technological changes in different fields into new forms of advanced technology.

FURTHER READING

Abernathy, W.J. (1978) *The Productivity Dilemma,* The Johns Hopkins University Press, Baltimore.

Elster, Jon, (1983) *Explaining Technical Change,* Cambridge University Press, Cambridge

Girifalco, L.A. (1991) *Dynamics of Technological Change,* Van Nostrand Reinhold, New York.

Kodama, F. (1991) *Analyzing Japanese High Technologies,* Pinter Publishers, London.

Nabseth, L. and Gray, G.F. (1974) *The Diffusion of New Industrial Processes,* Cambridge University Press, Cambridge.

Rothwell, R. and Zegveld, W. (1985) *Reindustrialization and Technology,* Longman Group Ltd, Harlow.

CHAPTER 3

The economics
of technology

3.1 INTRODUCTION

The management of technology needs to consider the economics of technology. Economics is concerned with items of value, their use and the choices that relate to them. As a subject it deals with one major aspect of human affairs. This chapter introduces those parts of economics which are particularly concerned with technology; it presents, so to speak, the relevant extracts. This involves selection and the drawing of boundaries. Readers who wish to acquire a fuller background are advised to consult such writers as Samuelson and Nordhaus (1992) or Lipsey (1989).

It is easy to observe the overall value of technology in ordinary life, but it is a challenge to measure it. The problem is its very nature as:

- technology has so many constituents, interactions and applications,
- some of its components are a public good, others are private assets,
- there is continuous change.

Considered at the micro-level, i.e. at the level of the firm, it is possible to measure the value of a technology. A common case here is the negotiated price for intellectual property, such as the licencing of a technology. It is also possible, for instance, to cost the use of alternative process technologies in manufacturing operations. This chapter will explore such economics at micro-level and introduce the concepts of technological economics and engineering economics. It will also present related concepts, such as production functions and the economy of scale. These are useful for the business management of technology.

At the macro-level it is much more difficult to assess the role of technology. Its contribution to national output is acknowledged; again the difficulty is the measurement of its amount. This is one reason why technology was for so long regarded as a 'residual'; statistically it could not be crystallized as, for example, numbers in employment. This has policy implications at national level.

3.2 THE MEANING OF TECHNOLOGICAL ECONOMICS

By its nature the subject of economics has a wide range of application and technological economics are interpreted here as economics applying to the use of technology. In essence, the economics of technology relate economic principles to the laws of science and, more specifically, to the practice of technology. There is a fusion between a social science and technology. This integration is essential for practical applications and decision making in many walks of life, but it has its problems.

While both require numeracy and sometimes entail sophisticated mathematical analysis, they are based on very different foundations. Technology is based on precisely defined units of measurement with a high degree of accuracy. The fundamental terms of economics lack such precision and empirical data are often difficult to obtain. With technology the distinction between the micro- and macro-levels is less explicit than it is in economic theory.

Technology is concerned with the practical applications of scientific principles – it is in the judgement of what is practicable that economics enter the decision context. Of course, it is not the only factor, but it is important.

3.3 EXAMPLES OF TECHNOLOGICAL ECONOMICS

Many of the standard textbooks for different technologies incorporate aspects of economic analysis, particularly where cost comparisons are made. The following are typical illustrations.

3.3.1 Electrical engineering

Kelvin's Economy Law is a simple but good example. The scientific component of the law concerns the resistance losses of electrical conductors. These are observable and measurable phenomena of nature. The economic aspects relate the operational costs of such losses to the original investment in the conductors. For instance, what is the most economic conductor cross-section for the distribution of electric power from a generating station? The task here is the minimization of costs when it is made up of two related variables. The following is an example:

let C = the total cost of a feeder line, p.a.
  C_x = the capital charges for the line, p.a
  C_y = the running costs of the line, p.a.

It can be shown that the capital cost of the line is a function of its copper content which can be expressed in terms of the line cross-sectional area 'a'.

Then $C_x = k_1 a$, where k_1 is a capital cost constant reflecting depreciation and interest charges. This is an economic measure. Similarly: $C_y = k_2/a$, where k_2 is a lost energy constant due to the resistance losses of the cable. These losses are inversely proportional to the cross-sectional area. This is a phenomenon of nature.

We can express the total annual cost function as follows:

$$C = C_x + C_y = k_1 a + k_2/a$$

differentiating with respect to 'a', we obtain

$$dC/da = k_1 - k_2/a^2 = 0 \text{ for a minimum value of 'a'}$$

solving this equation for 'a' we obtain

$$a = \sqrt{\frac{k_2}{k_1}}$$

This is the very simplest of relationships and in practice some very complex equations are used in the design of electrical distribution systems.

3.3.2 Engineering production

A very basic problem of technological economics concerns the speed at which to cut metals. The problem relates to the behaviour of metals of different hardness when forced against each other (science). The practice of production (technology) deals with the speed at which two metals move against each other, the rate of required penetration (feed) and the configuration of the harder metal (tool). The main physical characteristics of the operation are metal removal rates, heat generation, toolwear and surface finish.

In economic terms, the metal removal rate is the objective function. The higher the rate, the better, as it increases the rate of production and output. All other things equal, this also reduces the unit costs of the cutting operation. Heat generation is a complication and a cost is incurred to contain it. Toolwear is a cost, because the cutting tool has either to be replaced or be reground after a number of operations. Surface finish is a value, a condition of quality which is judged by fitness of purpose.

In practice, the problem reduces to the selection of a set of speeds and feeds in a turning or drilling operation which minimizes a total cost function. Consider a relatively simple and partial equation of the form:

$$VL^a = B/(f)^b$$

where the economic decision parameters are:

$$V = \text{cutting speed of operation}$$
$$f = \text{feed rate of tool}$$
$$L = \text{required drill life}$$

and the technological parameters are:

$$B = \text{constant depending on the operation carried out}$$
$$a, b = \text{indices depending on the material and the tool used.}$$

The technological parameter values are given for different metals with different hardness values. These have been derived from research investigations and are normally available in tabular or graphical form. Thus, quite complex computations are structured into a simple form. The structure of the equation denotes a trade-off decision; you increase one parameter value at the expense of the others.

3.4 THE SCOPE OF TECHNOLOGICAL ECONOMICS

The two illustrations indicate in a simple form the nature of technological economics. Examples can be taken from many different technologies and can range from simple cost-type formulae to complex mathematical analysis. They can usually be found in the technical publications of the field concerned and are often regarded there as little more than an appendix to the specialism involved. Generally speaking, phenomena, such as power losses, friction, wear etc., are major technological criteria; how we deal with them depends on economic aspects, such as: costs, time, interest rates, inflation forecasts. Again, economic dimensions feature prominently in trade-off studies, with the choice of different designs or processes. Engineering decisions are taken at the plant investment and at the operating levels. The former is concerned with the nature of the technology embodied in new equipment; the latter decides its pattern of use. At both levels technological economics are involved; cost data are set against output targets and contributions, efficiency and plant reliability requirements. It is appreciated that the principles involved cut across industrial and professional boundaries. For instance, the decision model for the optimization of catalyst life in chemical processes has a similar structure to that for the optimization of tool life in engineering production.

As a subject for the study technological economics could, typically, consist of the following:

- the economic measurement of technological attributes,
- the economics of technological change,

- the economics of research and development,
- capital/revenue relationships,
- the economics of scale,
- the economics of energy,
- the economics of manufacture,
- the economics of distribution systems,
- the economics of automation,
- the economics of transport systems,
- technology as an economic commodity.

A convenient range of economic analysis could relate to a selection of the headings represented by the first two digits of the Standard Industrial Classification (see Chapter 1). The chosen division or class could then be approached as shown in Fig. 3.1.

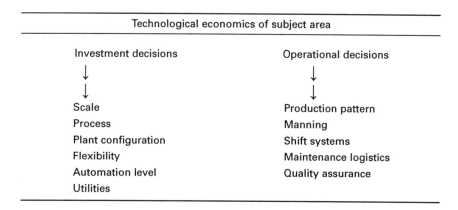

Fig. 3.1 A subject structure for technological economics.

3.5 ENGINEERING ECONOMICS

This subject area, also known as engineering economy, is a derivative of technological economics. It is mainly concerned with the economic aspects of the decisions made by engineers in their professional role. These apply, typically, to the design, manufacture and distribution of goods as well as the provision of services. Specialists in this field in the USA are known as engineering economists; their German equivalent the 'Wirtschaftsingeneure' are well established with an appropriate education to master's level at German technological universities.

Engineering economists are often concerned with the following decisions:

- which design to adopt,
- which equipment to choose,

- what process to use,
- what machine to repair or replace.

Essentially, these decisions are choices, involving the allocation of scarce resources. The data are expressed in economic terms such as costs, benefits, staff time, material usage. These data are derived from technological processes and characteristics.

Some of the most common forms of analysis are cost estimates and the derivation of cost functions using equipment and construction cost data. Although they are concerned with the economic properties of technology, they do not normally include technological parameters as such. Figure 3.2 illustrates the distinction between technological and engineering economics.

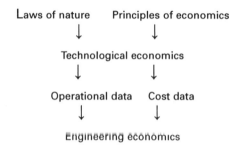

Fig. 3.2 The distinction between technological and engineering economics.

Engineering economics are concerned with such concepts as production functions, economic order quantities and learning curves. Typical techniques are cost/benefit analysis, break even computations, inventory analysis, plant investment appraisal, productivity measurement, value engineering, sensitivity and risk analyses. Much attention is given to aspects of financial management which impinge on decisions made by engineers. The subject is well covered by such writers as De Garmo (1979).

Essentially, the decision analysis is based on mathematical computation; the objective function is economic; the decision parameters are operational. In turn, the operational parameters are derived from the technological characteristics of processes and equipment.

A typical calculation in engineering economics is the derivation of the machine hour rate for major plant items in a manufacturing industry. An interesting example here is the standard VDI 3258, published by VDI (Verein Deutscher Ingeneure) the German association for professional engineers. Its recommended approach first establishes an annual machine cost rate which usually incorporates the following:

- depreciation costs, based on the estimated equipment replacement costs, not on depreciation for tax purposes. This focuses the mind on plant replacement reserves;

- interest costs, based on loan capital for half the equipment replacement costs;
- space costs, including light, heating, security costs;
- services costs, such as power and compressed air;
- maintenance costs.

The division of the total of these annual costs by the estimated hours used in a calendar year gives the hourly machine rate for production cost calculations. Actual calculations can be detailed and complex. Technological aspects are implicit in all of these categories.

In turn, such a computation can be applied to the choice of production processes. For example, Fig. 3.3 is a typical illustration of how an equipment choice can be made on the basis of unit assembly costs (including both line and labour costs) in relation to different rates of production.

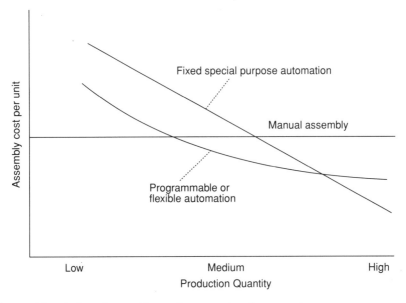

Fig 3.3 The choice of assembly equipment related to quantity.

3.6 PRODUCTION FUNCTIONS

These are an important tool of economic analysis. Typically for a manufacturing unit, the role of technology is indicated by the analysis of production functions which relate the physical inputs of capital and labour to a given level of output. Figure 3.4 shows an output isoquant QQ which can be achieved by different combinations of capital and labour. Although the diagram suggests a large number of possible combinations, only a small number of specific technologies are feasible in practice. These are shown by T where, for illustration:

T1 = capital intensive technology, such as full automation,
T2 = an intermediate technology,
T3 = labour intensive operation, with little technology.

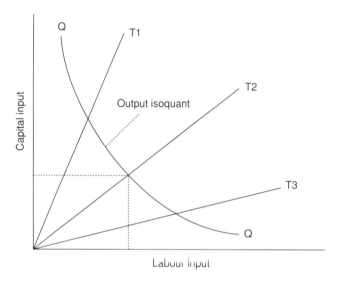

Fig. 3.4 The concept of a production function.

The physical capital input is the actual usage, the wear and tear, of equipment which is expressed in a provision for maintenance and depreciation. The labour input includes all operator and staff inputs which can be assigned to the output.

Figure 3.5 shows the effects of production efficiency improvements with a given technology T. The improvements can be attained by better deployment of staff, more effective equipment utilization, greater reliability, minor technical refinements etc. As a result, the same output can be achieved with reduced capital and labour inputs. The effect is shown by a move from line Q1Q1 to Q2Q2. Because of such practical variations in efficiency some economists prefer to view isoquants as bands rather than a line. The nature of these improvements is developed in Chapter 13.

Figure 3.6 illustrates the effect of technological change. Assume in the first instance that technology T1 is replaced by technology T2. As a result the isoquant Q1Q1 moves to the parallel position Q2Q2 but retains the same value. The same proportions of inputs remain and in that sense the technological change is neutral. The economic gain of the change is shown by the reduction of the required inputs for the same production quantity.

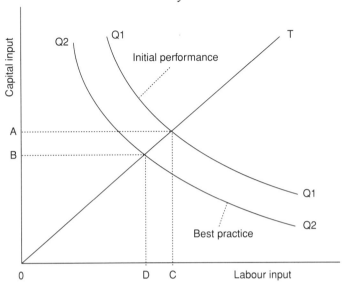

Fig. 3.5 Efficiency improvement with a given technology.

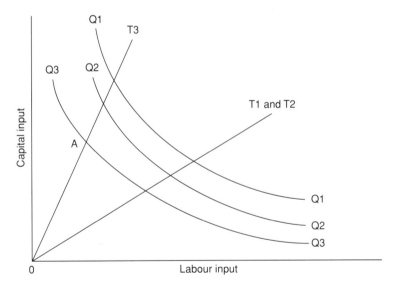

Fig 3.6 The effect of technological change.

Progress generally is expressed by the production isoquant moving towards the origin of the diagram.

The technology line T3 indicates biased technological change as expressed by line Q3Q3. Bias means a different mix of labour and capital inputs. The nature of the bias or direction of the technological change should be noted. For instance, point 'A' can express the installation of an automated production line which is capital intensive with a high degree of embodied technology. The labour content of production is correspondingly reduced.

3.7 THE CONCEPT OF ECONOMY OF SCALE

This is an important concept for the management of technology. The concept is concerned with size and its implications, such as structure, complexity and costs. Economy of scale can be visualized at the following levels:

1. a physical unit, such as a distillation column or a production line, for instance for electronic assembly,
2. a factory, consisting of one or several production units,
3. a business, with one or several factories,
4. an industry, consisting of a number of firms,
5. the national level.

Level 1 is most clearly linked to a given technology, as the scale ascends, managerial and organizational aspects become more important. Economy of scale is a feature of importance with industrial operations and is particularly associated with capital intensive manufacture. It can be defined as the reduction in unit costs as the size of an entity increases. Size can be expressed in various physical terms, manufacturing capacity, number of employees and actual output. Economy of scale can apply to the following activities:

1. research and development, including preproduction,
2. plant construction, machinery and equipment,
3. plant operations, including management of continuous operations,
4. marketing, sales organization and distribution;
5. technology, sales, control techniques, application knowledge, softwareprograms.

Economy of scale presumes an environment, a set of given technologies and input factors, such as the labour content of a particular manufacturing process. If these alter, as with the introduction of a new technology, scale effects can change. It is important to specify at what level an economy of scale is to be determined. For instance, at level 1 there can be strong economies of scale based on technological factors. This does not necessarily provide a case for economies of scale at level 3 where a large corporation becomes cumbersome because of organizational problems.

The most important activities affected by the economy of scale are manufacturing operations, where it is relevant both on capital and on operating account. The reduction in unit capital costs may be due to technical and shape factors, particularly in the process industries where, for instance, with storage or reactor vessels, capacity is a function of volume and cost a function of surface area. Where capacity is expressed in terms of proprietary units, such as with packaging machinery in the food industries, the 'quantity discount' opportunities of commercial negotiations also offer some reduction in unit capital cost. Similarly, there are operating economies of scale, such as with manning levels or with unit maintenance spares inventories. Such benefits are additional to the advantages of large-scale operation, such as bulk buying, distribution and the utilization of by-products. These depend, of course, on the volume of available business.

It is possible to express the expense function of a manufacturing plant, in a general, composite form which integrates various design and operating parameters.

Typically, scale effects can be of the form:

$$E(v) = K(v)^x$$

where

v = plant size (in output units)
$E(v)$ = capital cost of plant with capacity 'v'
x = economy of scale factor, where $x = \leqslant 1$
K = constant

The lower the value of x, the greater the economy of scale and the cheaper will be the plant per unit size. In some process industries, such as petrochemicals, the values of x can be as low as 0.6, while in some branches of the engineering industry, with discrete production, it is difficult to obtain values of x below 0.9. When $x = 1$ no benefits accrue from the scale factor. Diseconomies of scale arise when $x > 1$; i.e. the plant or operations are too big to achieve minimum costs. Figure 3.7 shows the scale effect on unit costsfor a production plant, with a scale factor $x = 0.6$. Other things being equal, a doubling of size will reduce unit costs by nearly 25%.

Economies of scope are a related concept to the economy of scale. They refer to the application of standard components, products and processes to a range of uses. Other things being equal, wider application increases volume and this yields economy of scale. Economies of scope arise from the flexibility of manufacturing processes and from generic products, such as chips for computers.

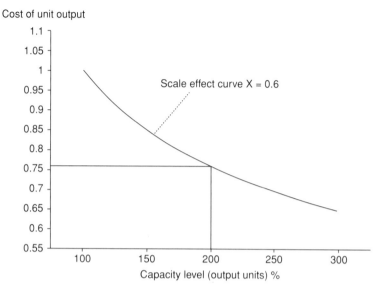

Cost of unit output

Fig. 3.7 The effect on unit costs with scale factor x = 0.6.

3.8 THE CONCEPT OF OPTIMUM SIZE

It is an accepted principle of economics that an operational unit, whether in manufacture or distribution, is regarded to be of optimum size when, other things being equal, it can operate at minimum costs. Smaller or larger units incur additional costs. Where a unit is too large, diseconomies of scale are incurred. It is also well known that the optimum size of a unit can vary with industry and technology. Size depends on the technology used and on market characteristics. One of the effects of technological change can be a change of the optimum size. Consider, for instance, a machine shop where traditional equipment has been replaced by computer-controlled machine tools. As these require no significant set-up times; the economics of batch sizes have been transformed. Where previously the optimum batch size could have been, say, 250 is is now one unit. Moreover the flexibility of the equipment and its unattended operation generate much higher efficiencies and a much smaller machine shop yields the same or a greater output. It will, of course, be appreciated that the optimum size of an operating unit is not necessarily the same as that of a business.

With continuous incremental change and learning experience in both construction and operation, the optimum size of production units in some of the process industries has changed dramatically. For instance, in the petrochemical industry the capacity of a typical ethylene plant in 1952 was about 30 000 tons per annum (tpa). By the early 1980s plants of a capacity of 500 000 tpa were coming on stream.

Economy of scale can be as much a matter of distribution as of production. For instance, in the manufacture of glass or plastic containers there are considerable economies of scale in having large units of production but the diseconomies of scale for transport costs of moving high volume, low unit value goods tends to counter this. The optimum size for these two cost patterns together is a medium size unit with a carefully chosen location to contain transport costs to customers.

3.9 TECHNOLOGY AS A COMMODITY

The most familiar transfer of technology is in an 'embodied' form; i.e. it is incorporated in fully developed products or processes. It can also be transferred as intellectual property in the form of patent rights, by licence agreements or as key know-how. In such cases the licensee may carry out detailed product and manufacturing development. Technology, as an intellectual property, is a marketable commodity, the value of which is governed by the economic expectations of prospective users. As such it is subject to supply and demand, albeit in an imperfect market. The nature of technology transfer is considered in more detail in Chapter 12.

Generally, technology is marketable in the following forms:

1. Patents. The assignment of patent rights through licence agreements. A licensee obtains the use of specific inventions.
2. Registered designs. These apply particularly to consider goods where, in addition to function, appearance and style are important product attributes.
3. Design right. This provides protection for an overall design which is mainly based on function rather than style.
4. Copyright. While this is better known for the protection of literary works it applies also to drawings and computer records.
5. Trade mark. While not a direct dimension of technology it can give expression to the quality and excellence of technology associated with a product.
6. Know-how. This category of knowledge has market value because it encapsulates relevant experience such as commissioning skills, control techniques, application knowledge or software programs. While much of this finds its way into manuals and guide-lines, some knowledge is difficult to document. This is process skill and judgement is often person-specific. According to Lawrenson (1993) know-how extends beyond the technical and manufacturing functions to marketing, finance and other enabling knowledge appropriate to a particular technology.

3.9.1 Innovation as an economic process

According to this concept an economy no longer adjusts passively to technology but becomes the instrument for determining the extent, the nature and the articulation through time of the development of a technology. Technology provides the opportunity for what Schumpeter (1961) describes as 'new combinations' which form the basis of economic development. Entrepreneurs are those who use these opportunities to create new products, methods of production, new markets. Innovation is the means through which such new combinations can be found and developed. The technical and commercial quest for innovation is one of the hallmarks of enterprise which makes a major contribution to economic progress.

3.10 TECHNOLOGY AT THE MACRO-ECONOMIC LEVEL

Essentially, the relevant economics of technology here are those at the micro-level, i.e. at the level of the firm or industry. However, for the sake of completeness, a brief reference will be made to the analytical treatment of technology at the macro-economic level.

There is little doubt that much of the increased standard of living enjoyed by most industrial and industrializing countries has been based on improved and new technologies. These have resulted in:

- growth in the productivity of an economic system,
- the introduction of entirely new goods and services,
- the enhancement of existing goods and services.

The neoclassical theory of economic growth, as expressed by an overall production function, explained growth in terms of two factors of production: capital and labour, as typified by the form of the Cobb Douglas Function (Cobb and Douglas, 1928)

$$Q = Q(K, L)^1$$

where Q is the aggregate output, K and L are capital and labour inputs. Cobb and Douglas took the statistics for the US economy as fixed capital, total capital and labour supply published by the US Census Bureau and

---

[1] The equation will be of the form

$$Q = A\,K^{\alpha}\,L^{\beta}$$

where $\quad$ A = constant
$\quad\quad$ α $\quad$ β = coefficients for capital and labour respectively

the National Bureau of Labor Statistics. Cobb and Douglas used the data in index form to derive regression coefficients and constants. In broad terms, the relationship between the coefficients expressed a set of technologies. If the relationship changed this could be due to several factors, not only technological change. However, the economy was seen to grow because of the general diffusion of technology.

The tremendous, visible technological changes taking place during their main period of investigation, 1899–1922, first suggested the incorporation of a time function 't' which would indicate the contribution to national economic growth by technological progress. The basic form of the macro-economic production function, as expressed by Solow (1957), then became:

$$Q = Q\,(K, L, t)$$

Since then a sophisticated framework of econometrics has been built on this basis. The contribution of technology to economic progress was seen as the main explanation between the observed and the predicted values of the time series. It explained much of the 'residual', i.e. the errors of the regression analyses.

Measures of aggregate output, such as the Gross National Product (GNP) do not fully reflect the benefits of technology and innovation. This is particularly so where an innovation is directed to the final consumer. For instance, the development of power tools for domestic purposes, such as DIY (do-it-yourself) substantially increases productivity in the home – but this is not measured.

There are also cases where the GNP can be diminished by technological progress. The example of the improved reliability of aircraft engines illustrates this. Because of continuous development and product improvement the reliability of aircraft engines has substantially improved; an engine can now remain in place for 10 000 flying hours. As a result, airline operators need fewer spares. For the engine manufacturer this has reduced an important line of business; the GNP is reduced accordingly although the reliability of airline service has improved.

3.11 SUMMARY

Technological economics relate the principles of economics to the laws of science and the practice of technology. The subject is relevant both to investment and operational decisions. Its derivative, engineering economics, is concerned with the economic aspects of the decisions made by engineers on the basis of relevant cost data. Technology itself, as a commodity, is subject to supply and demand, albeit in an imperfect market.

Production functions, as a tool of economic analysis, indicate how the capital and labour inputs to production are affected by technology. Economy of scale is an important feature of industrial operations, particularly with capital intensive manufacture.

FURTHER READING

De Garmo, P. E. *et al.* (1979) *Engineering Economy*, 6th edn, Macmillan Publishing Inc, New York.

Institution of Civil Engineers, (1969) *An Introduction to Engineering Economics*, Institution of Civil Engineers, London.

Lipsey, R.G., (1989) *An Introduction to Positive Economics*, 7th edn, Weidenfeld and Nicolson, London.

Pratten, C.F. (1971) *Economies of Scale in Manufacturing Industry*, Cambridge University Press, Cambridge.

Samuelson, P.A. and Nordhaus, W.D. (1992) *Economics*, 14th edn, McGraw-Hill, Inc., New York.

White, J.A. *et al.* (1977) *Principles of Engineering Economic Analysis*, John Wiley & Sons, New York.

Technology and Management

CHAPTER 4

Corporate technology strategy

4.1 INTRODUCTION

Part Two of this book concentrates on the management of technology. To begin with, this chapter looks at the basic intent of a business and how technology features within this. It has the following objectives:

1. to show the relationships between business intent and technology,
2. to demonstrate the need for a corporate technology strategy,
3. to describe the requisites for, and the nature of, such a strategy.

Readers of the financial press and business reports will note that successful enterprises are often explicit about their intent. This is contained in, or linked with, a mission statement which encapsulates the main purpose, the *raison d'être* of the business. Such a statement of intent is the point of departure for the development of business strategy and policies.

Portfolios of technology are inherent in manufactured products and the processes by which they are produced. Technological opportunities provide the scope for competitive advantage. Technological capability allows a company to take these opportunities. This capability is normally firm-specific, cumulative and path dependent. It is the outcome of strategic decisions. Technology is the basis for added value and a high added value is a major contributor to profitability. There is a need, therefore, for a corporate technology strategy.

As Twiss (1990) points out, the purpose of a technological strategy is to provide a stream of goods and services to further the company's business objectives in a changing world. The nature of advanced technologies, the rate of technological change and the associated pressures of competition argue for such a strategy. This chapter considers the character and the implications of such a strategy. It applies not only to companies which wish to achieve and to maintain world-class status but to any significant industrial company which wishes to survive. The setting for strategy within business direction is indicated by Fig. 4.1.

Corporate technology strategy

Fig 4.1 The setting for strategy.

4.2 THE BUSINESS MISSION

Overall business strategy starts with a mission statement which sets the basic purpose of the firm. To be effective this has to be a vision, a succinct declaration which forms the anchor point of all the goals and policies to come. The mission statement specifies in overall terms who the customers are and what market sector they constitute; the benefits provided by the enterprise and how they are furnished.

4.2.1 Corporate goals

A good mission statement is terse and direct. The specification of goals is a development and amplification of such a statement. Without goals there is no focus and without focus there is less motivation. The statement of goals needs to be explicit and compatible with the types of activity that staff can plan.

In the context of technology it is appropriate to state the goals in terms of applications of core product and process technologies. According to Rouse (1992) sales and profits then become the consequences of the mission, not the goals. The goals become the basis for strategy development, planning and implementation.

4.3 WHERE IS THE BUSINESS?

The formulation of strategy starts from the given position of a business. The assessment of the business position contains the following major tasks:

1. The evaluation of the business. This is an operational audit of the nature of the business, its technologies, the strengths and weaknesses

of its products and processes, the capabilities it can muster. Much depends on the markets and the general context in which it operates and the stage of growth it has reached. For instance, Dodgson (1989) indicates the following distinct stages of business growth:

- start-up,
- technological and scientific consolidation,
- internationalization of markets,
- professionalization of management,
- vertical integration,
- product and business diversification.

2. The assessment of the technological environment. Technology assessment is the first step in the audit of a company's position relative to the technologies it currently uses or develops. The next stage is to forecast how external technological developments affect the activities of the company. Technology forecasting contains a group of specialist techniques which will be discussed in Chapter 5.
3. The comparison of the company with its competitors. Evaluation requires comparison. The company needs to establish how its performance relates to the average and the best performers in its field of business. One way of doing this is by benchmarking.

4.3.1 Strategic benchmarking

Benchmarking is a systematic process for the evaluation of and comparison with companies recognized as industry leaders. It can be with one particular firm or a group of companies. One well known international electronics corporation benchmarked itself against 30 leading firms in its various business areas. There are no geographic limits here. It is a search for best practice, wherever that might be found. This is more than an 'intelligence activity'; it is a basis for strategy and action.

Benchmarking can be both at the strategic and operational levels. At the strategic level it starts with published company information; particularly the accounts of the company to be assessed. The overall standard management ratios can usually be derived without too much trouble. What return on capital? How do profits relate to total sales revenue? What are the time-series for key ratios? While in many cases such information may be readily available, some of the data that really matter are not. First of all, the accounts for many large companies are consolidated and give little away about specific activities. Secondly, little about the use of technology can be extracted from company financial statements.

More often the products and services of the industry leaders are the first focus of attention. The speed of respective product launches can be compared. Product portfolios and attributes can be assessed and performance specifications can be obtained. It is also often feasible to apply 'reverse engineering' to the products of the industry leaders; to test and strip them, to examine components, to assess methods of manufacture and assembly and, by value analysis, to cost them. This can reveal the technical and design strengths and weaknesses in competitive products. It is also important to carry out systematic market research; to listen to the customers who have the choice of products. Again, much can be gleaned from the views expressed by key suppliers. A typical product benchmark profile is shown in Fig. 4.2.

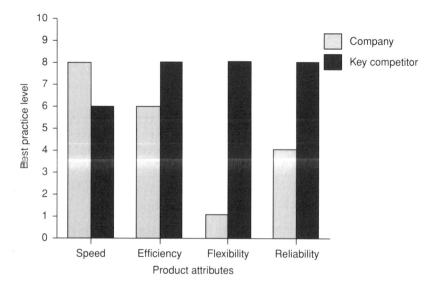

Fig. 4.2 Product benchmarking against key competitor.

Other indicators, such as added value per employee, or output per operator are also significant. Crucial ratios, such as the proportion of direct to indirect staff or engineers for product development, as compared to manufacturing development, give important leads about the strategies of competitors.

Benchmarking is not just a snapshot of a given situation. From a strategic view the longer-term aspects of benchmarking are also important. Much can be learned from patent searches. These give some indication of what is to come. The frequency of patenting, the subject matter, the detailed techniques, the claims all give an insight.

Benchmarking can also be applied to companies considered for acquisition as part of a strategic plan. This is a particular challenge where the target company does not wish to be acquired. For instance, to relate such a business to world-class standards needs an evaluation of its products, customer base, equipment technologies, capabilities, prospects, management style – quite apart from the normal financial investigations.

Benchmarking is a regular but intermittent activity which needs to be carried out systematically. Its yield is a measure of the gap between a company and that of the industry leaders at world-class level. With it should come:

- an analysis of the 'success factors' that could account for the observed differences,
- a set of feasible targets for strategy formulation.

Another form of benchmarking is inter-firm comparison, usually facilitated by a clearing house or professional organization. This can be at strategic or operational levels where it is particularly useful with non-competitive partner companies in different sectors who use generic technologies and have comparative systems. A whole hierarchy of activities and indicators, such as engineering changes or warranty claims can be compared. Significant differences and resulting discussions are a stimulus and innovation rich.

There is a need to consider also inter-firm and network possibilities. A network is a well established set of relationships between the firm, its main customers and its key suppliers. The density of innovative transactions within such networks can be examined.

4.3.2 Technological gap analysis

It is possible for a firm to estimate its prospective market and profit position on the basis of the two scenarios: with and without innovation. If the company ceases to develop new processes and products its market share and sales will gradually decline, even with the further market promotion of existing products. Compared with continued innovation, a market and profit gap will develop which broadly indicates the opportunity costs of not innovating. Fig. 4.3 illustrates the outcome.

This form of gap analysis is particularly important in industries with a high rate of innovation, such as microelectronics. The gap in sales is one aspect; the gap in profits can be even more important. It is not only the question of a performance gap but also of an opportunity gap. 'Why did we not grasp these opportunities?' Without innovation the company has to compete in what are often mature products and markets with tight margins, whereas the leadership with new products may realize better rates of profit.

A firm would be ill-advised if it develops its technological strategy in a vacuum. It needs to understand the strategy, technical position and the achievements of its competitors. How is their know-how generated, what are their strengths and weaknesses? What are their prospects? It is useful to have indicative benchmark projection of key variables such as cost and product performance trends, shown in Figs 4.4 and 4.5.

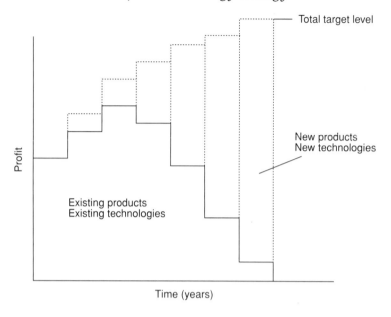

Fig 4.3 Contribution of new products/technologies to profit plan.

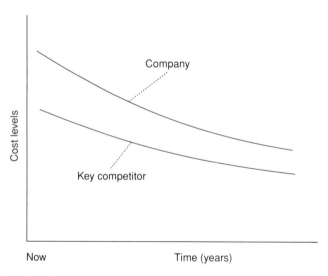

Fig 4.4 Benchmark projection of cost levels.

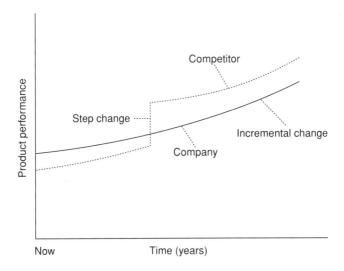

Fig 4.5 Benchmark projection of product performance.

Technological forecasting is a first step towards closing the gap. How it is to be closed is a major decision; whether incremental changes are sufficient or a step change is required.

4.4 THE CONCEPT OF BUSINESS STRATEGY

Strategy is the disposition and arrangement of resources to achieve business objectives in a competitive market setting. These resources must be commensurate with the intent. Essentially, they comprise groups of human, physical and financial resources. The development of strategy is a complex, adaptive process. In a changing environment it enables a company to steer a course that allows it to get ahead of its competitors. The formulation and periodic review of strategy allows a company to adjust such a course, if necessary, in due time and without undue costs. It is important to have a carefully managed flexibility in response to external events, without sacrifice of the business mission.

The business strategy will encompass all functions and levels of operation and when fully developed it will spell out the relevant components in a manner which is meaningful to those in the positions concerned. Particularly with an innovation strategy, all aspects of business operations are affected. If there is insufficient integration of strategies, then the overall intent becomes unbalanced and the risk of failure is increased. The strategy will be sufficiently specific and integrated to assist managers to choose between alternatives. It will be expressed in terms of time and money.

Strategy is concerned with the anticipation of innovative possibilities and the assessment of changes in the business environment. An example of this is new markets with different types of customers for a newly developed product.

4.4.1 Competitive strategy

Porter (1985) sees competitive strategy in the following manner:

> If a firm can achieve sustainable cost leadership (cost focus) or differentiation (differentiation focus) in its segment and the segment is structurally attractive, then the focuser will be an above-average performer in its industry.

In essence then, business strategy aims to achieve maximum competitive advantage for the company. The benefits – however enumerated – should follow.

Competitive advantage must be expressed in quantifiable terms. Cost leadership is to be the lowest-cost producer in an industry. This allows the firm to tap those segments of a market where there is a high price elasticity of demand, which in turn can sustain a product volume with significant economies of scale. Differentiation is achieved by clearly defined product groups which focus and concentrate on specific market segments. It can be protected by patents, registered designs or brand loyalty.

A company needs to set its objectives in relation to the best of its industry. The setting of goals which are essentially incremental to past performance is inward looking. It takes no account of the gap between the firm and the best in its industry and it is that against which it has to compete on a world-wide basis.

In the development of strategy, customers and markets need to be envisaged first. For instance, Siemens AG, the multinational German corporation, which stresses this approach, has set up its business units to match them. The business divisions are organized to be flexible and they each have their own R&D. The need for this view can be appreciated in terms of some of its key markets such as electronic components, energy, health care, organization systems, materials, traffic and communications.

4.4.2 The classification of strategies

There are a number of ways in which different types of strategy can be classified. In this context the division by Urban and Hauser (1980) into initiative and responding strategies is the most useful. The following summary illustrates their approach:

(a) Initiative strategies

1. The strategy uses R&D based innovations resulting from in-house development.
2. The strategy is mainly entrepreneurial. The innovation activity is opportunistic but not necessarily technologically novel.
3. The acquisition of products or companies making the products.
4. A market-based strategy using opportunities for competitive innovation.

(b) Responding strategies

1. The firm reacts to customer requests for innovation; e.g. the supply of electronic components in packaging which allow direct feed into automatic component placing machines.
2. The strategy is imitative, reacting to the new product introduction by a competitor.
3. The 'second-but-better' approach, developing and improving on a competitor's innovation.
4. The defensive strategy. This may be contained to modifications of existing products to remain competitive. R&D is undertaken as a precaution against surprises from competitors.

4.4.3 Strategic change

No strategy can be absolutely fixed if the world around the business changes, but adjustment requires careful consideration. As Ansoff (1979) points out, strategic change is not an instantaneous event, but a protracted time and cost producing process. The first step in the process is the search for a new product/market combination, which as Jacobsson (1986) observes is usually both expensive and time consuming. The scope of the search is generally affected by the affinity between the target areas and those in which the company currently operates.

A strategic change can be one or more of the following:

1. diversification of the company's products and markets,
2. expansion of the company's operations,
3. transformation of the company's operations,
4. the adoption or abandonment of a major technology.

With an expansion strategy many of the current resources employed by the firm will be similar to those required for the expansion, whereas with a diversification strategy, new skills, know-how, and possibly different forms of organization will be needed. These constitute a greater risk than those associated with a pure expansion strategy. The third category of strategic change is often associated with companies facing severe competition and a low return on assets. They have to change to stay in business. The change often has to be fundamental in terms of product and manufacturing rationalization as well as in the style of management and the culture of the firm. Technological change can trigger such a process.

4.5 THE CAPABILITY FOR STRATEGIC PLANNING

The proper development and the periodic review of a business strategy requires considerable staff support. This includes the preparation for the next-generation technology. It needs the capability to react rapidly and effectively to technological change. A strategy team will include the technology, marketing, manufacturing, personnel and finance functions. The skills and resources needed for strategic planning have to be assessed and marshalled. This may call for considerable preparation and training. Also, the integration of the planning activities is important and the interface between staff specialists or consultants and top management requires careful definition.

The report of an Institution of Electrical Engineers expert mission (1993) 'Semiconductor lasers and packaging: an insight into the Japanese optoelectronics industry' makes an interesting reference to the research planning departments of several major Japanese companies. Typically, they consist of up to 20 staff who control the research direction, strategy and the finance of the total R&D activity. The planning departments are run by senior engineers, supported by younger scientists seconded for about 1–2 years. They also include accountants, personnel and intellectual property rights (IPR) specialists. Generally, they report to a board consisting of divisional business managers and R&D directors.

4.5.1 A structure for strategic planning

Rouse (1992) emphasizes the need for a structured approach to strategies for innovation. If there is a structure the focus is on the content of the strategy; if there is no structure then there will be a preoccupation with structural matters. The structure of strategic planning is seen as a series of steps. It starts with audit – Where are we now? How well are we doing? What are our expectations and assumptions about the future? The training stage is to ensure that strategic planning is carried out in a professional and time-effective manner. The prework stage is a period of formative investigation and dialogue so that sets of scenarios and proposals can be put to strategy meetings. The decisions made at such meetings, and such thinking that is appropriate to it, must be transmitted and explained to all concerned with the implementation of the strategy. This communication stage is followed by action plans.

A useful expression of a structured approach has been developed by Danila (1989). In his strategic formulation of high technology projects he uses the framework of a 'support graph'. This has three stages. The first is concerned with the definition of objectives and strategic alternatives. The thinking is divergent and consists of identification, conception and formulation. The second stage is convergent and deals with the feasibility, evaluation and decisions about the global and sub-objectives. The third stage concentrates on the realization and control of the strategies. It identifies key tasks and the human, technological and financial resources required to achieve them.

Unless a major external event calls for an immediate response, the updating of strategy requires periodic reviews, typically on an annual basis. One of the most important criteria of such an analysis is the added value for customers from company products.

An interesting example of strategic planning, quoted by Turner (1993), are the competitiveness achievement plans (CAP) introduced by Lucas Industries. The first emphasis is on the identification of customer needs. This includes the analysis of technology trends and the evaluation of both customers and competitors. The key industry success factors are established and competitors are benchmarked on these. From this the business objectives are set for the three-year CAP period and operational plans developed.

4.5.2 Organizational aspects of strategy

One of the risks of a changing business environment is that organizations can be encumbered by their founding ideologies and past successes, which stress competences that are no longer in tune with current challenges. Established managerial values and industry-specific beliefs can generate particular approaches to decision making and the role played by formal analytical and structural processes. Ways of embedded thinking and routines can result in a discriminating approach to innovation opportunities. The internal company political setting is seldom neutral to the prospect of change.

The strategic decision structure of a company is significant. At board level financial and marketing issues are the key decision subjects. Unless there is a substantial investment request there is the temptation to leave R&D and technology matters to the specialists who, because of their organizational position, do not always have the opportunity for the development of a strategic view.

In a sense, strategic management is concerned with future-oriented pattern recognition. It requires clear vision, direction and flexibility. Pettigrew (1977) views it as a process of building awareness, momentum and commitment to change among managers and groups within the organization. Strategy then emerges from a complex political process of demand generation and mobilization of power around these demands within the organization.

It must also be appreciated that different functional groups within an organization have different views and attitudes towards technology. There is the important role played by the champion.

4.6 CORPORATE TECHNOLOGY STRATEGY

On the basis that the technologies employed by a company are an important group of resources we can extend Porter's concept of strategy. A corporate technology strategy is therefore concerned with the use, development or adoption of technologies to maximize the competitive advantage of the business. We have already noted that the value of a technology is best assessed in terms of competitive advantage. Technology yields this if it can improve the company's relative cost position or product differentiation. In this light a firm has to consider which technologies to develop; what risks to take; whether to be a leader or a follower in a particular technology. Often there is no ready-made choice of technology, only of logical progressive paths which could lead to them.

To achieve competitive advantage, the technology strategy must be integrated within the company's overall business strategy. This means it makes a contribution to the overall intent but also gets support from it. There is no room for splendid isolation here! A divergence weakens the overall strategy and the hoped-for competitive advantage may not be realized. This does not only apply to the choice of a technology but also to the level at which the technology is pitched. Sophistication, not valued by the market, is at best an irrelevance. If it brings its own complications then it diminishes competitive advantage.

The form in which a technology can make its contribution can, of course, vary. In addition to cost reduction and product value there are organizational benefits. An increased operational, particularly manufacturing, flexibility will enable a firm to take more opportunities in the market place. Also better control and integration of activities, such as computer-based information systems will increase the speed of response to market opportunities.

One important aspect of a technology strategy is the systematic exploration of emerging technologies. By this is meant a well-defined, structured approach of assessment which judges both the promise and the relevance of such technologies to the business. In a major group such analysis may be carried out at a corporate rather than a divisional research centre, particularly if an emerging technology has generic potential, such as advanced materials. The adoption of an emerging technology or, for that matter, any technological change merely because it is an advance has no justification as such.

4.6.1 Formulating a technology strategy

A strategy is developed through a number of systematic tasks. Porter (1985) gives a succinct overview of the tasks that need to be undertaken. The company has to:

1. identify all the distinct technologies used by the business,
2. identify potentially relevant technologies in other industries or under scientific development,
3. determine the likely path of key technologies,
4. determine which technologies and potential technological changes are most significant for competitive advantage,
5. assess the firm's technological capability in those technologies and the costs of making improvements,
6. select a technology strategy which integrates the selected important technologies to reinforce the company's competitive position.

Where there is a large corporation with a number of business units he recommends the following:

• reinforce business unit technology,
• identify core technologies that impact on many units and fund these to establish a viable level of knowledge and staff,
• ensure coordinated research efforts by the business units,
• ensure the diffusion of results to all business units,
• use acquisitions, joint ventures or know-how agreements to introduce new technologies.

Other relevant requirements are suggested by Maidique and Patch (1978) who stress the assessment and improvement of technological competence, appropriate investment and staffing of the R&D function, and competitive timing; i.e. an initiative versus a response strategy.

The determination of a technology strategy calls for plans and projects and the allocation of resources to them. How is this to be done? One useful approach is that of Jantsch (1967) who quotes the practice of project selection on the basis of resource matrices.

Each project proposition shows the required tasks and scientific disciplines in one dimension. The company resources are expressed in terms of scientific manpower, physical facilities and available finance as the other dimension. The projects are then aggregated to show the resource implications for the company if it wishes to realize a proposed strategy. In a similar manner research resources, which are an indicator of technological strength, can be related to market prospects.

The importance of this type of matrix analysis is in its structure of thinking and the type of questions it prompts, particularly about the end uses of a new technology. The analysis can be used both for yes/no decisions as well as for other strategies, such as the forming of consortia or the purchase of technologies.

Another relevant approach is the multiple contingency concept of long-range technological planning developed by Lea and Miller (1968). This uses a projection technique to integrate technology forecasts into business plans by:

- projecting as many plausible scenarios as possible,
- developing contingency R&D plans for each scenario,
- assessing how the numbers of R&D projects change with the number of scenarios considered,
- planning R&D projects around the areas of greatest overlap.

4.7 COMPETITIVE TECHNOLOGY

It is a matter of strategy for a company not only to select a technology but also to pitch it at a competitive level. Competitive technology for a firm consists of levels of technology, properly developed, managed and effectively used, which, with given resources, provide the maximum competitive advantage to a company. Furthermore, as a company rarely uses just a single technology, a competitive portfolio of technologies integrates such technologies to achieve maximum competitive advantage. With given resources this will involve the setting of priorities in technology development.

There is no permanence with a given competitive technology. Also, there is no competitive advantage with a technology that is available to all competitors in an industry. As a company enhances its resources, particularly in terms of the skills of its staff, the level of technology which maximizes competitive advantage and added value will also change. Indeed, the enhancement of resources is a business objective which can lead to further growth in added value. Competitive technology is not necessarily the highest level of technology, particularly if that is not well managed. That could be, and has been with some companies, 'a technology too far'.

The following are the main features of a competitive technology:

- other things equal, it maximizes added value,
- it enhances the competitive advantage of the products, which incorporate it,
- if it is a core technology, associated technologies are well integrated with it,
- it can be effectively used by the staff of the company,
- it is reliable,
- it has a limited life in the market place.

4.7.1 Technology as a resource

In the absolute sense no company could ever be self-sufficient in the technologies which it applies. Whether a manufacturing company or a service organization, it will use many technologies embodied in the equipment required for its operations. The use of technology in such a proprietary form is not an issue.

However, there are firm-specific technologies which are integrated in the goods and services which the company provides. Self-sufficiency, coupled with differentiation, is one way of safeguarding the company's long-term competitiveness. There are core technologies which are vital for the very existence of the business, such as engine technology for a car manufacturer. Some are shared technologies to that firm, such as paint and finishing processes for car bodies using automated, high volume equipment. The firm is likely to share technological application know-how with paint manufacturers, the key suppliers who formulate and develop paints to agreed specifications. The car manufacturer is also the user of proprietary technologies, incorporated, say, in electric motors for machine drives. Although it is possible to categorize technologies in terms of usage, control or emphasis, technical or commercial changes can, of course, affect such a categorization.

An alternative classification has been put forward by Little (1981). The first group are the base technologies which are extensively used in a given business. For instance, in a clothing firm, such technologies refer to the cutting, sewing and garment assembly operations. They are used by all firms in the industry and, as such, do not confer a competitive advantage. Key technologies are those which for the time being confer competitive advantage, such as computer-controlled garment cutting. The acquisition and exploitation of such a key technology, which reduces the cost and increases the flexibility of manufacture, would, typically, be an objective of technology strategy. Pacing technologies are those in the development stage where the current applications are limited yet the potential for competitive advantage is promising, such as the application of adhesives for automatic garment assembly. This classification can be used to provide the following guidelines for the investment in technology:

Technology	*Approach*
Emerging	Monitor
Pacing	Invest selectively
Key	Develop and control
Base	Divest selectively

This classification applies to the criterion of competitive advantage. Dussauge, Hart and Ramanantson (1992) suggest that what is a base technology in one industry could be a key technology in another.

With the explosion of technology in the last 40 years it is becoming impossible for a firm to master all the core technologies which it needs for its operations. It must therefore depend on the flow of ideas, know-how, patents and other intellectual property to keep up to date.

4.7.2. Technology development

This is primarily concerned with the development of those technological capabilities which a company needs to meet business objectives within a period of, say, ten years. It comprises both the strengthening of existing capabilities and the introduction of entirely new ones. It is a strategic process which includes the choice as to which technologies should be developed and which left alone. Where a company commits itself to the development of a specific technology, then it has to decide how and where this technology should be acquired and funded. Where it is a core technology it may develop the technology further within the firm; with a shared technology it may form a partnership with a specialist supplier. With a proprietary technology it may use its role as a customer to encourage technological development by component manufacturers

Technology development consists of a range of activities concerned with the improvement and introduction of products and processes. This is a much wider concept than research and development. For instance, a manufacturing quality assurance department evolves a more effective statistical process control with the use of spreadsheets on personal computers. This results in higher product reliability. In the eyes of the customer this can be as much a value enhancement as a design improvement.

4.7.3 Technology assessment

This can either be an independent review or linked with technology gap analysis and benchmarking. It is the regular review of the company's technological strengths and weaknesses in relation to its products and processes. These have to be related to current and future business contexts. A review of the technological prospects of the company includes the identification and assessment of technological opportunities and options. The implications of such reviews have to be considered, such as the acquisition of new technologies, future skills requirements, recruitment and training needs. Some of the techniques of technology assessment are outlined in Chapter 5.

4.8 THE FOCUS OF STRATEGY

4.8.1 Concentration on core business and core technology

Comparative advantage lies in the ability of firms to use their technological and managerial skills, experience and know-how in a specific area. This is normally based on a cluster of core or key technologies. If the portfolio of company activities significantly exceeds these then the business may become over-extended. Financial expansion, on its own, is not sufficient.

Corporate key technologies are the fundamental base for technological competition. The life cycle of a key technology may span a number of successive product life cycles and can often be measured in terms of decades. It is part of corporate strategy to spell out clearly what these key technologies are. For instance, one major electrical corporation has defined the following as some of its core technologies: Energy, electronic components, network and systems, software and materials. Advanced materials technology was recently adopted because of the crucial role of materials in the manufacture of sophisticated electronic components and the difficulty of procuring these to the required specifications. In operational terms each core technology was the R&D responsibility of specific business units. Again, a privatised UK water company, as part of a major business change plan, has defined its areas of engineering excellence as environment, infrastructure and process engineering, integrated with a core competence of managing water and waste water systems.

It will be appreciated that progress in a specific core technology can depend on achievements in a range of related technologies. Also interlinkages between various technologies have to be considered.

However, Twiss and Goodridge (1989) indicate that the dominant position of a company's core technology, as expressed by its senior technical staff, can sometimes discourage the recruitment of high calibre staff in a new technology. As a result, there could be an imbalance in design effectiveness, such as poor reliability of electronic systems incorporated in mechanical engineering products.

Firms may decide to maintain a technological capacity without corresponding production as a hedge against the tactics of competitors and an otherwise uncertain future. Protective R&D will generate technologies which will only be marketed when the company is threatened by a rival.

4.8.2. Technology leadership

As part of his general thesis Porter (1985) indicates that technology leadership becomes valuable where it enables a firm to gain a cost or market advantage. Particularly with new products the firm has the ability to set up a basis for competition, such as channels for distribution and product standards; it can shape much of its commercial context.

However, where there is rapid technological change, the continuing quality of the company technology becomes crucial. Also, with rapid product improvements incorporated in new models or updates, continued funding is important to maintain the flow of such products. The timing of product launches and updates becomes critical not only to get ahead but also to stay ahead of competition.

The attractions as well as the risks of technological leadership are well illustrated by an article in *The Times* of 30 April 1980. Headed 'Bitter lesson of the body scanner' it reported on the Thorn-EMI decision to withdraw from the medical electronics business. The bitter lesson was the cost of the world lead in new product technology. Here the brain and body scanners which, with the combination of X-ray and computer processing technologies, had been a giant step forward in the techniques of medical diagnosis. The reader was reminded that there is no natural law that such a genuine technological breakthrough would automatically result in profitable business. The company lead was challenged by manufacturing problems. In marketing terms the company saw its main opportunity in North America. Although it did not have a suitable marketing infrastructure it decided to enter the market on its own, although with hindsight it might have been better to have linked with an American company familiar with the requirements of a large but exacting market.

The scope of the product breakthrough was such that there were many more orders for scanners than could be produced in the required time. The scanners were plagued with development problems; by the time these were overcome much of the market advantage was lost. There were insufficient production engineering skills to focus on scanner manufacture, resulting in delivery times of about a year compared to the four months offered by quick-learning competitors. This was at the time when many hospitals of substance in the USA were willing, and able, to purchase scanners. The opportunity to establish the company's market position was transient. The finance required to sustain scanner operations until production problems had been solved was no longer available. The opportunity was perceived but its implications had not been fully appreciated. Technological and market leadership was lost.

4.8.3. The maintenance of technology leadership

Porter also considers how technology leadership, once attained, could be maintained. In his view this depended on the following:

Firstly, the source of technological change. If the change comes from within the firm or the industry it will be more controllable. If it comes from outside the industry it will be more difficult to sustain technological leadership because there is the risk of general access to the change. In such a case, where a technological advance is crucial for entry to the industry the barriers to entry will be correspondingly reduced.

Secondly, the existence of a sustainable cost or differentiation advantage due to the technology development activity. As a technology matures the emphasis will change from product differentiation to cost leadership.

Thirdly, the relative technological skills. Where a firm has the staff with superior technological skills it will be easier to maintain its lead position.

Finally, the rate of technology diffusion. This depends in turn on the awareness of competitors of the generated technological changes. This is where technological evolution, rather than a major breakthrough, attracts less attention. It also depends on the resultant scale of benefits competitors envisage and their capacity to take advantage of the changes.

The strategy embraces all the technologies which form, or could form, part of a company's operations, not just the R&D programme. For example, the introduction of a computer-based information system in a marketing department can improve the speed of company response to changing market conditions or, with inventory control, it can contribute to cost reduction. This is particularly important where advances in generic technologies have a potential impact on company operations. As an example, bar coding is well established for volume distribution of goods in supermarkets. Similar opportunities exist in production control and the general movement of components within an engineering factory. Such an opportunity – and its attendant costs and problems – are not normally part of the R&D programme of a potential user company. The risks are not that the opportunity is necessarily overlooked but that it is not given the same level of rigorous scrutiny in policy terms that it may warrant.

4.8.4. The technology-cluster strategy

This is a particular strategy which is based on the development and strength of generic core technologies. The hallmark of a generic technology is that it can be combined with other technologies to give application opportunities in diverse markets. A prime example is telecommunications, which has applications in satellite systems, interactive video, smart cards, telenewspapers, home banking, electronic fund transfer, ID verification, fax systems – to name just a few.

A company may carry out substantial research in its own laboratories or with universities and research associations to develop a carefully chosen basis of advanced technology. Advanced materials technology and microelectronics have been suitable areas for such developments. The technologies can also be acquired from outside the company. In such cases the company will concentrate intensively on its further development until it is the leader in the field. In either case the company needs a strong scientific and technological capability.

A cluster of related technologies provides a company with many application opportunities in diverse industries. The strength in the cluster technologies is then combined with an organizational structure and management philosophy which spurs on the rapid development of new products and markets for these technologies. Such a strategy enables the firm to operate several businesses, linked by the generic technologies which provide what are known as 'diversification pivots'. The businesses of the company are defined by their technological potential and are governed by technology application opportunities. The company is able to move quickly into an industry and when its technological advantage has been eroded, it can quickly move out of it. The firm competes more with its technologies than its products.

4.8.5. The integration of technology and patent strategies

In strict commercial terms the yield of patents is extremely variable. There are instances of private investors having made a fortune from an opportune and usually simple invention. There are also critical patents which, especially in the pharmaceutical industry, have been the basis of great commercial success. Yet, studies quoted by von Hippel (1990) suggest that for most industries patent protection is not seen as useful for excluding imitators or capturing royalty income. The very publication of a patent specification gives competitors intelligence about the research progress achieved by a firm and the likely innovations to come on the market. The skill of filing protective claims for a patent is matched by the expertise of circumvention. The response to infringement is often expensive and troublesome and many firms are deterred from this unless the stakes are really high.

Nevertheless, even where circumvention is likely, there is a time lag to it and with short product lives the time advantage by the patent holder is significant. Also, circumvention takes R&D resources and it may be more profitable to both parties to come to an agreement on the use of a patent.

The purchase of a patent right is one thing, but that does not necessarily give a firm the know-how and experience to use it effectively. One way is to involve the firm in joint research which generates such patents.

4.8.6 The link with manufacturing strategy

The integration of technology strategy with the overall business strategy of a company is particularly relevant to manufacturing strategy. The body scanner case already mentioned shows how technology leadership can be negated when the need for a corresponding manufacturing strategy is not appreciated at the critical stage of product development. For an industrial company there has to be a symbiosis, a mutual reinforcement, between the two strategies. The appropriate deployment of manufacturing facilities is part of the competitive strategy to achieve increased market shares. This subject matter is further developed in Chapter 7.

4.8.7 The strategy of application engineering

This is a particular strategy often used by equipment and component manufacturers. It is essentially the identification, the concentration on and the technological solution of customer problems where a product has to be tailored to suit specific application requirements. It may involve considerable customer liaison, design and test effort at the customer product design stage. Such a strategy can be staff-intensive and calls for careful engineering management. On the other hand, it provides the opportunity of new market niches as well as intelligence of market changes. It was in this manner that, for example, a UK plain bearing manufacturer, specializing in engine bearings, became a major supplier of structural bearings for motorway bridges.

4.9 TECHNOLOGICAL ALLIANCES

4.9.1 Combined research and development

In recent years the costs, risks and time-scales of substantial R&D programmes have escalated so much that even major companies have to consider alliances to reduce the financial burden and still provide access to advanced technology.

There are a number of reasons for this. For instance, social and environmental changes have a substantial influence on the technology strategy of some industrial companies. As an example, the motor industry has to respond to the demands of environmental legislation and safety requirements. While this reflects the social requirement of the 1990s, its expression varies with different countries and can involve minute details of legislation. Simultaneously, rapid technical changes in car electronics, engine management systems and new materials have pushed companies into new development programmes which are continuously becoming more expensive.

The design and development of a new volume production car in 1992 cost at least £200 million and this heavy level of required expenditure has induced about a dozen European car manufacturers to form the Prometheus consortium, thus combining competition in the market place with technological cooperation.

Another significant partnership development has been the recent agreement between IBM, Toshiba and Siemens to work together on the

256-bit electronic chip. These are major, world-class companies, all with substantial financial resources. Yet the challenge is so great that a techno-logical alliance is seen as the way ahead. Each company has specialists in different aspects of research and their constructive interactions are expected to provide synergy to the project.

The heavy costs of developing high thermal efficiency gas turbines for power generation has prompted the establishment of global part-nerships or consortia. The key suppliers are transnationals with divi-sions across the spectrum of related business and manufacturing loca-tions in many major world markets. For instance, GEC Alsthom is the expression of a working partnership of its British and French parent companies. It designs, supplies and builds complete fossil-fired power stations. Its other divisions supply diesel engines, boilers, power trans-mission and distribution systems, electronic controls and much of the related plant. Until the 1970s these were designed and supplied by sep-arate companies. Although a leader in the field of steam and gas tur-bines, it has formed a partnership with General Electric (GE) of America to transfer and develop know-how from aircraft engineer usage to power station applications.

Another form of alliance, increasingly encouraged by governments, is the pre competitive research consortium, such as the British Alvey Programme (1983–88). This was an example of research cooperation with substantial government financial support and which included no less than 113 firms and 55 university departments. Such a programme avoids duplication of research effort, reduces risks and makes better use of specialist staff. A partner company is less likely to end up with a 'marooned technology'; that is out on a limb in the market place. A useful alliance has leverage in joint capability without overlap in the respective core businesses. Following the completion of the programme member firms are free to compete. Alternatively, an alliance can be extended to manufacture and marketing, such as with the European Airbus Consortium.

4.9.2 Technology exchange agreements

Technology exchange agreements are often appropriate where firms in the same industry operate in different world markets. Each company has its own R&D programme but, on the basis of the exchange agree-ment, it may concentrate on different aspects of product and process research. Such agreements also provide an opportunity for the transfer of specific skills and expertise. It is important for a company, as part of its general negotiations, to identify in some detail its skill resources and to endeavour to close its skill gaps as part of the bargain. The exchange can relate to the cross-licensing of patents and the mutual disclosure of know-how.

In some cases the relationships between the members of the alliance can become so close as to establish control of the market. This is achieved by cross-licensing, know-how agreements and the confidential exchange of information.

4.10 THE REALIZATION OF STRATEGY

Corporate technology strategy will specify, within a given horizon, targets for the development of technological advances as well as the acquisition and exploitation of new technologies. It will be integrated with the overall corporate plans and take into account human, financial and physical resources. The corresponding task programme, together with the authorized budgets and resources, will become the basis for implementation and subsequent assessment of performance.

A strategy also needs to be understood and appreciated in an appropriate form at other levels of the organization. Uncertainty about strategy inhibits initiative, the taking of opportunities and long-term planning. One major UK utility company tackled the implementation problem of a major new strategy by forming 16 project groups, each with a nucleus of a director and three senior managers.

Apart from the general implementation of strategy at the various operational levels of the organization there are a number of specific aspects which relate to its realization.

4.10.1 The technological breakthrough opportunity

One of the benefits of effective technology forecasting and opportunity orientation is to spot an opening for a technological breakthrough. This is seen by Ozawa (1974) as a substantial and early return on a relatively modest R&D investment. Typically, this is possible with a successful technology application or where development leads to a substantial enhancement of a key performance parameter, e.g. a reduction in engine fuel consumption. As technological progress frequently has no linear relationship to development expenditure, it can occur when a technological barrier has been overcome. The occasion for this is often the development and transfer of supporting technologies from other industries.

The perception of an opportunity depends on how it is defined. Its assessment therefore requires:

- the identification of the key performance parameters in economic and market terms,
- the definition of the degree of parameter enhancement which is feasible and promising,

- a cost estimate of the R&D required to achieve the target enhancement,
- the time-scale of realization in technological and economic terms.

4.10.2 The development of expertise into new business units

A successful technology strategy and corresponding R&D programme at times yields new business opportunities in other technologies. The acquired expertise can be harnessed to new market opportunities and their realization also requires a strategic approach.

Malpas (1987), a former managing director of British Petroleum International plc showed how key technologies derived from the company's R&D can generate new businesses. The role of BP Ventures was to develop new businesses even if these did not align with existing operations. For instance, subsidiary companies were concerned with composite materials, photovoltaics, special flares, boilers, separators and energy management. All these ventures were expected to become profitable and some to become substantial. Those which eventually did not become material to the core activities of the company were likely to be sold. A profitable sale was regarded as a successful venture.

Another illustration of the opportunities resulting from a carefully developed technology strategy is that of a well known chocolate confection manufacturer. The firm had successfully automated most of its production lines and as part of this policy had developed a strong capability in automation, electronic process control and precision equipment manufacture. Consequently it found new opportunities in the development of vending machines and coin counting equipment which now form the key products of a new business unit. This is a form of spin-off where a developed technological capability finds an outlet in new business applications.

A further example is the volume production of plain bearings for the automotive industry. A major supplier found standard machine tools inadequate for high volume bearing production and this encouraged the company to develop its own special-purpose machinery. The design and supply of such specialized plant has, in turn, developed from a manufacturing service into a substantial separate business with major export contracts and a significant licensing income.

4.10.3 The integration of the supply chain

The relationships between customer and supplier are obviously a function of scale and transaction volume. In some industrial sectors, such as the motor industry, the total procurement costs and the attendant purchasing expenses are a substantial part of the total manufacturing costs. The competitive cost advantage is therefore significantly affected by acquisition costs. This is not just securing a supply contract at the lowest price but covers the total cost of managing or, better still, working with a supplier. It involves a limited number of first-tier suppliers, where trust and close collaboration can lead to optimum performance in terms of cost, quality, delivery, responsiveness and customer satisfaction. There is also the opportunity of harnessing the best of the technological resources of the suppliers, particularly where these are firms of substance. Where there is significant technological change, long-term product development programmes can be integrated and key staff can work side by side from the earliest stages of a new design, starting with a joint component specification. The development of such relationships between independent commercial organizations needs care, time and transparency.

The choice of suppliers reflects company strategy based on continuous improvement and the adoption of new technology. The spectrum of relationships is not confined to firms selling raw materials or components but now includes suppliers of services and computer software. Strategic sourcing aims to secure a firm's competitive advantage by the development of an effective supply system. This means the establishment of committed suppliers, capable of contributing to process and product innovation, quality, timeliness and cost control.

The basis of the relationship is a 'win–win' approach; that it is to the long-term advantage of each party to work together and not be mesmerized by a short-term gain. Although the theme may be mutual profitability this does not prevent the setting of ambitious targets, such as with cost reduction.

The selection of a supplier is on the basis of vendor analysis and certification, typically on the basis of the International Standards Organisation ISO 9000 standard. It presumes that a supplier already has or is committed to develop a total quality management (TQM) system and culture.

Vendor analysis is concerned with the current performance and the overall capability of the supplier who is expected to move at the same rate in technological development and training. The quality of supplier staff and the control of their selection is an important condition for close cooperation, particularly as the integration of design activities is becoming a major facet of supplier relationships. This includes participation in the concurrent engineering of products and processes, joint process studies and problem solving.

Where the company has adopted an overall computer-based management system the suppliers must be able to be integrated within an electronic data interchange (EDI) system. This is particularly important for delivery schedules and just-in-time deliveries within a context of continuous flow management to the company's assembly lines. In some industries bar coding has also become a condition of supply.

Some leading companies no longer regard vendor certification as a one-off exercise. The relationship between the company and a key supplier is the responsibility of named staff, not necessarily a procurement manager. He is associated with joint forward planning, quality improvement projects and, in some cases, training programmes.

Despite the close cooperation, assessment of the supplier continues. Rating procedures are developed and rating scores are reviewed jointly. Where necessary, improvement plans are agreed and followed up. Such substantial investments in key supplier relationships presume a corresponding level of business. The investment is based on mutual interest but is not dependent on sole sourcing.

Suppliers in Japan play a much larger role in carrying out the detailed engineering of individual components. According to Jones (1992), some of the major Japanese car manufacturers base their commercial relationships with suppliers on a joint analysis of costs and setting prices on an agreed profit level. The need for the supplier to make a reasonable profit is acknowledged in a win–win relationship which also includes a formula for sharing the gains of cost reduction over the life cycle of the product.

Such a basis of mutual cooperation is fundamental where the customer operates a flexible manufacturing system, relying on the continuous delivery of small, zero defect batches. For this, meaningful and reliable data transmission both ways is critical. Also, the integration of design activities is becoming a major facet of supplier relationship. This includes participation in the concurrent engineering of products and processes.

4.10.4 Relationships with dominant suppliers

There are other cases where the suppliers of particular equipment or systems are powerful corporations with leading edge technology and the user companies are small/medium size, operating in a competitive field. The realities of business relationships may not enable the relatively modest firm to insist on its requirements at an affordable price. It may here be more appropriate to design its products to match the supplier's equipment. In most cases such equipment, e.g. electronic controllers, is supported by the established technology of the major supplier and is best used as a self-contained module. The supplier, in turn, may pay attention to the technological progressiveness of the customer as well as the volume of business. As has been noted, for instance, in the field of instrumentation, new product ideas often come from the problems of the modest customer.

4.10.5 Technology and the international corporation

One important facet of technology strategy is where in the world it will be implemented. The international corporations, 'the multinationals', have particular opportunities for technology development and flexibility of technology deployment. They can carry out and control R&D in various parts of the world to correspond to the global spectrum of their resources and to hone both technology and product design to national or regional market requirements. Problems with different national legislation, customs and practices, such as with directives by the European Union, can be resolved within the countries concerned.

Recent years have seen a steady globalization of research. For instance, DRT International (1991) reports that an increasing proportion of the research by multinational companies is no longer carried out in the home country. As an example, it quotes the German corporations, Bayer and Hoechst, spending 36.9% and 40% respectively of their 1989 research budgets abroad. Such location diversification is fostered by the need to be close to important markets, particularly where product adaptations are needed. Significantly too, is the local availability of educated staff and skilled labour. There could also be tax incentives, such as enhancement of deduction or credit against tax liability for R&D performed locally.

There has also been an increased movement across national borders of key staff within these corporations. For instance, *The Times* of 26 February 1992 refers to the use of contract research and technical services within the Ford Motor Company and its opportunities for technology transfer. This is enhanced by a policy of 'co-location' of engineers from the UK and Germany so that those employed on particular projects work side by side. There is also considerable interchange of engineers between Dunton and Cologne to improve efficiency.

One of the organizational problems with multinational companies is to resolve the degree of local autonomy with which local business units can operate. Is there to be global uniformity or local diversity? The practice of Philips International BV is of interest here. Its diverse businesses are divided into divisions, such as consumer electronics, components etc. For further technical development, particularly of processes, one factory serves as the divisional lead unit. In that sense technological initiative is mainly decentralized. Even then there is further diversity depending on the location of different factories. For instance, with the manufacture of television tuners, which are components of considerable complexity, a European factory has a much higher degree of automation than a sister factory in South East Asia, with its lower unit labour costs. The emphasis on process improvement differs with the economics of production.

4.10.6 The challenge to smaller companies

The technology strategy of the smaller company is governed by limited staff and financial resources. This challenges the attainment and maintenance of technology leadership. If the firm wishes to retain its commercial independence it has the following typical choices:

- concentrate on part of a product or system, e.g. compressors for refrigerators,
- focus on a niche market where it can achieve a substantial market share,
- form a partnership with a similar size company.

4.11 THE TECHNOLOGY CRISIS

However thorough the development of a company's strategy might be, there is the risk of an unexpected technological development which completely cuts across the company intent. What should a firm do in such a case? According to De Greene (1982) technological crises should not be viewed as undesirable but as fundamental elements of the continuity-and-change syndrome. In a genuinely dynamic mode of behaviour, change is natural: technological momentum, not inertia, creates security. Strategists need to tolerate the uncertain and be disposed to learn experimentally.

Despite the negative connotations of the term 'crisis' it can be a galvanizing event and a learning opportunity. According to Loveridge and Pitt (1990), innovative opportunities can become crises when a firm either does nothing or adopts inappropriate strategies. Crises can arise where the changes predicted by the firm fail commercially. This may be due to the unexpected recovery of a defender technology, the insufficient appreciation of the potential of a novel concept or mistaking the dominant trend of a new development.

If there is insufficient appreciation of the nature and extent of a crisis then there is the temptation to make minor adjustments and to retain the continuity of existing technical and administrative routines. This can result in what Johnson (1988) calls a form of 'strategic drift' which, in turn, can lead to a more serious form of crisis.

As Ford (1988) points out, strong challenges and technological positions are often based on 'bundles' of technologies. Early signals of such impending challenges may not be deemed significant until they form a convergent pattern.

It must be appreciated that crisis are perceptual phenomena; events emerge as crises because they are seen as such by members of an organization. The analytical response may be affected by bias, overconfidence and reliance on insufficient evidence. There is also the problem that a committed view offers resistance to subsequent non-confirming information.

4.12 SUMMARY

Technology strategy is an essential and integral part of industrial business strategy. The formulation of strategy starts from a given business position. Benchmarking and trend projections relate the company to its competitors. Technological gap analysis indicates the effects of continuous innovation on company prospects. A company needs to have both structure and organization for strategic planning. Achievement is based both on the formulation of a technology strategy and its effective implementation.

Competitive advantage is supported by competitive technology. This consists of levels of technology, properly developed, managed and effectively used, which, with given resources, provide the maximum competitive advantage. Corporate key technologies are the basis for technology leadership. However, the increasing costs and risks with advanced technology R&D foster technological alliances.

FURTHER READING

Ansoff, I.H. (1979) *Strategic Management*, Macmillan Publishers, London.

Dussauge, P. Hart, S. and Ramanantsoa, B. (1992) *Strategic Technology Management*, John Wiley & Sons, Chichester.

Loveridge, R. and Pitt, M. (eds) (1990) *The Strategic Management of Technological Innovation*, John Wiley & Sons, Ltd, Chichester.

Porter, M.E. (1985) *Competitive Advantage*, Collier Macmillan Publishers Ltd, London.

Rouse, W.B. (1992) *Strategies for Innovation*, John Wiley & Sons Inc., New York.

Twiss, B. and Goodridge, M. (1989) *Managing Technology for Competitive Advantage*, Pitman Publishing, London.

Wild, R. (ed) (1990), *Technology and Management*, Cassell Educational Ltd, London.

CHAPTER 5

Analysis for technology strategy

5.1 INTRODUCTION

The purpose of this chapter is to specify and to describe the main techniques of analysis required for the development, direction and review of technology strategy. These techniques can be classified as technology assessment and technological forecasting.

Chapter 1 stressed that corporate technology strategy is part of business strategy and needs to be fully integrated with it. Technology strategy cannot be developed in a vacuum. Management needs a basis for judgement and intuition for this is insufficient. A systematic set of tasks needs to be carried out to gather, structure and audit information; to relate this to the company and its prospects. Then it is necessary to come to a view: to make recommendations.

These activities are management techniques and, basically, staff tasks. In larger companies a director normally has the special responsibility for planning and strategy. The remit may be linked to other responsibilities – this is a matter of scale and division of labour. Suffice it in the first place that someone is responsible for it. In the simplest case the person charged with it carries it out personally but the more likely situation is one where the detailed work is carried out either by management services staff at corporate headquarters or by a team of line managers and functional specialists seconded for the task. To make effective use of these techniques requires substantial information and the very task of seeking it fosters consultation.

Where a team is involved it is important that a top manager forms part of it. At director level he has better access to an overall information system and this helps to steer the activities of the group. Where decisions about primary research projects have to be taken, a consultant, such as a university professor, can often make an important contribution. Also, as the tasks involved are expensive in staff resources a framework of project control is needed.

5.2 TECHNOLOGY ASSESSMENT

The review of a company's activities includes the technologies currently incorporated in its products and operations. The latter comprises both the process technologies, such as in engineering production, and the associated staff activities, such as computer-aided design. Technologies come with assumptions and it is part of technology assessment to separate these from the technologies themselves. This requires a well structured approach and searching questions.

Technology assessment has two basic constituents.

1. the review or audit of the technologies employed,
2. a comparison on a benchmark basis between these and the best industrial practice.

The assessment team should include a senior engineer from an operating unit, a research manager and a corporate planner. This should ensure that three basic judgements are focusing on the technologies employed.

The objectives of the assessment are:

- to describe and verify the technologies in use,
- to evaluate the cost and the added value of the technologies employed,
- to identify the strengths and weaknesses of the company's technological operations,
- to indicate how the competitive advantage of the company could be enhanced by a better use of the present technologies,
- to identify available technologies which the company could incorporate in its products and operations,
- to establish the impact of and the added value from the use of new technologies,
- to assess the technological options available to the company.

Technology assessment is essentially a staff task but, working closely with top and line management, the assessors are normally expected to make recommendations how new or improved technologies can best be introduced and integrated within the operations of the company.

The technical review can help significantly in the identification of technological opportunities within the business boundaries and across them. The following are typical areas of application:

- energy conservation
- yield improvement of process plant
- maintenance reduction
- information technology

Technometric assessments can be used to measure the technical attributes of various products or processes, such as the efficiency of compressors or the power factor losses of a factory electric supply system. These can then be integrated into indices of technical performance.

5.2.1 The definition of strategic technical areas (STA)

According to Mitchell (1988) a 'technical area' can be viewed as a bundle of skills or a discipline which are applied to a particular product or service that addresses a specific market need. He quotes as examples lithographic techniques, the fabrication of semiconductor integrated circuits etc. A technical area is defined as strategic when it becomes critical to the survival or growth of a company.

For the purpose of analysis the STAs can be listed as a column in a matrix table with the company products in a row as shown in Fig. 5.1.

STA	Product			
	1	2	3	4
Manufacture of circuit board	0	0	1	1
Flow soldering	2	3	3	2
Automatic assembly	2	1	1	0
Automatic testing	2	2	3	2

Technology level
0 = not used
1 = lagging
2 = average
3 = leading

Fig. 5.1 STA matrix for electronics manufacture.

It will be noted that a given product may have a number of STAs; conversely a particular STA can apply to a wide range of products. In turn, the company's capability in the various STAs is benchmarked in relation to its competitors, typically under the heading of 'leading, average or lagging'.

When the STA profile has been established for the company it can use this to get a better measure of any particular area. This may only require a modest input of company staff resources. It can then make a strategic decision such as concentrating on the development of a specified technical area or deciding to contract out a given process to specialist companies.

5.2.2 The identification of technological impact

A particular new technology, such as optic fibres, can have a range of impacts on the products and processes of a manufacturing company and it is therefore important to define the affected areas. For example, as part of a project for the Singer company, Lyon and Rydz (1981) developed a table where the various company functions, such as product planning, application engineering and manufacture, constituted the vertical axis and different disciplines, such as graphics, simulation, mechanics, human factors and reliability typically formed the horizontal axis. The entries on the function column indicated those areas that had the greatest potential for interaction with the new technology. The entry along the horizontal axis indicated the professional skills that were required by the affected activities.

The assessment groups consisted of staff at operational level from relevant functions and disciplines but also included outside experts. They evaluated the new technology on operations within the company as well as on US industry as a whole. The assessment groups rated the new technology in terms of the following criteria:

- potential impact on activity
- timeliness of payback
- ease of skill acquisition.

The assessment groups were linked with a management review group which considered and rated the fit of the new technology with the corporate business plan.

The company had considerable success in correctly forecasting the impact of technological change but the commercial outcomes of some of its resultant ventures were mixed. One important factor with the less successful ventures was the lack of sufficient marketing inputs.

The setting up of interfunctional/interdisciplinary groups on a matrix basis had a number of important implications. The choice of assessors and their interaction during their deliberations brought some good product application ideas to the surface. This form of decentralized assessment drew from a wider pool of experience and observation. Secondly, where a project emerged from the assessment, there was a group ownership and commitment to the development. Thirdly, the relationships between the assessment and the management review groups were important, particularly as far as the steering and directing roles of the latter were concerned.

In a wider sense, Lyon and Rydz stress that technology assessment is not a once for all activity. It is best carried out annually as part of a company strategy review. The assessment focuses on the technologies most relevant to the achievement of company objectives. Key areas are divided into a number of general topics to ensure that general trends are brought to the attention of operating divisions. Specific technologies are then discussed in more detail and a forecast of their progress made.

5.2.3 Strength and weakness analysis

The purpose of technology assessment goes beyond the establishment of the company position. Current capabilities have to be linked to operational objectives. This has two aspects. The first is to establish what technologies have to introduced, developed or improved to meet these objectives. The implications of this have to be worked out. The second is to reconcile these objectives to the implications.

It is basic to the analysis that the related financial and marketing aspects to be included. One convenient method of relating these are by a two dimensional 3×3 or a three dimensional $3 \times 3 \times 3$ matrix approach. The analyst can use such matrices to explore all aspects of a particular prospect. For instance, Fig. 5.2 relates market strength to technological capability.

Market strength		Technological capability		
	S	Buy technology	Develop technological capability	Concentrate on opportunity
	A	Keep out	Look for opportunities	Strengthen marketing function
	W	Keep out	Find niche	Look for partners
		W	A	S

Legend
W = weak
A = average
S = strong

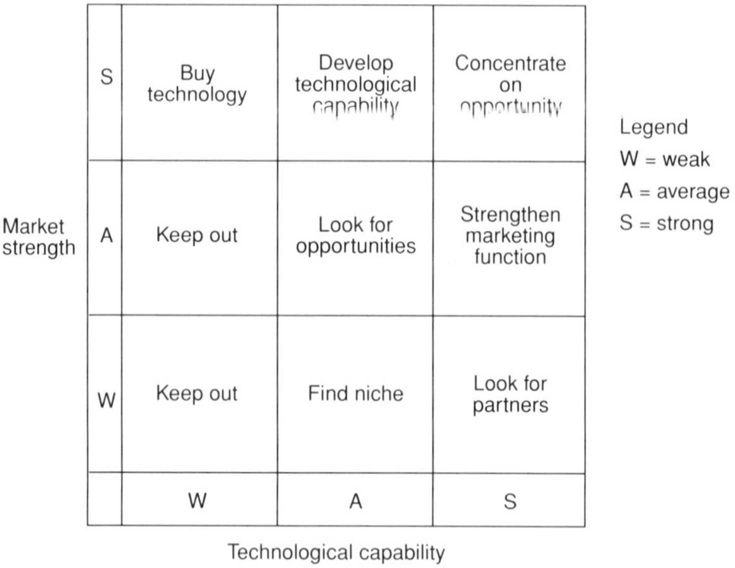

Fig 5.2 Matrix of market strength and technological capability.

A variant of this compares technological uncertainty and business exposure. The first expresses the probability of meeting technological performance objectives within a given period and budget. The other spells out the effect on the business if the objectives are not met. In the same way competitive advantage can be related to technological risk, or any other set of parameters can be chosen. Although most of these judgements are subjective, a matrix makes these more explicit.

5.3 TECHNOLOGICAL FORECASTING

There is no precise boundary between technology assessment and technological forecasting. In practice, the former has stronger links with the operations of a company and a relatively short-term horizon, whereas the latter with a longer time scale is more concerned with the development of strategy. Freeman (1987) suggests that technological forecasting within a well-established technological system is relatively straightforward, but it is much more difficult when there are structural changes. Then the critical task is to identify the main features of a new paradigm and its potential. The distinction affects the staff that carry out these tasks.

In essence, a forecast is a prediction of how a given variable will change within a stated period. This is linked with a probability estimate of the anticipated result. Technological forecasting (TF) consists of a set of formalized processes for the study of future technological developments due to advances in science and changes in society. Jantsch (1967) defines it as the probabilistic assessment, on a relatively high confidence level, of future technological transfer. Cetron (1969) puts technological forecasting into a more specific form. He defines it as a prediction, with a level of confidence, of a technological achievement in a given time frame with a specified level of support. Alternatively, a date can be assigned when an event has, say, a 50% chance of coming about.

There are two branches to technological forecasting. The first, exploratory or technological opportunity forecasting (TOF), starts from established knowledge and assesses the future. It considers how current developments came about, where they may lead to and the opportunities this will offer. The observed trends can be projected by simple extrapolation, 'S' curve fitting or trend correlation analysis.

The second, normative technological forecasting, takes a future goal or agreed problem and works backward to the current state of knowledge. It is often expressed in mission terms, such as getting man on Mars. It would specify the technologies required for the sustenance of life on the planet, indicate the direction of research required to achieve it and estimate a time-scale for its possible realization. On a broader scale, it establishes the future needs of societies and the resulting market potentials. It enumerates all possible goals affecting a chosen field of technology. The relative importance of those goals is then assessed by weighting or ranking together with the probability of achieving each of them. Again, the technique works backwards to ascertain the technological developments that are needed to satisfy these. This approach has become known as technology demand forecasting (TDF). A good example here is the need to develop drugs to neutralize the antibodies caused within the recipient of organ transplants.

Three time-scales for technological forecasting can be established. The long-term forecast, with a horizon of over five years, is concerned with macro-economic and social developments and the evaluation of scientific potential relevant to the goals of the business. It can contribute to the formulation of company policy. A medium term forecast, for a period of two to five years, will concentrate on market and technology application research and can assist with the definition of strategy. A short-term forecast, for a period of less than two years, will focus on details of products and market potential, with particular attention to the position of competitors.

Where an existing technology faces the challenge of a new one technology substitution analysis can be used to predict the rate of transformation. The classic illustration of this is the replacement of sailing ships by steamers, but more recent illustrations, such as the replacement of electro-mechanical devices by microprocessors can also be cited. Once a new technology has been established the pattern of substitution often follows a 'S' curve when the rate of fractional substitution is proportional to the old technology to be replaced.

Twiss and Goodridge (1989) suggest that an appropriate time-scale needs to be established for the analysis of the future. This should be based on a broad judgement of the likely developments of the main technologies which could affect the firm and its markets. The judgement would have to take into account the rate of technology advance and the speed of response to developments. It requires the identification of social, economic and political factors, their likely interactions and the impact they will have on the structure of the industry and the markets it serves. It will be important to establish not only what is likely to happen but when these developments can be expected and what their financial implications will be.

In recent years there has been a strong Japanese concentration on technology forecasting and the setting of detailed technical objectives, the achievement of which would be financed irrespective of short-term economic problems which the firm might face. This represents a technology push situation and often reflects strong technology representation on company boards.

5.3.1 The role of technological forecasting

As technology is a major instrument of global competition, technological forecasting is one preliminary step towards competitive advantage. To this end, it facilitates the evaluation of:

- the market potential for a particular product or process which incorporates new technologies,
- the rate of progress for a particular technology,
- the impact of new technologies on a company or industry.

Within the firm, technological forecasting is a constituent of the company research planning system. The objectives of that system are:

- to establish meaningful research objectives,
- to make the company aware of major long-term threats and opportunities,
- to develop the research component of the overall company strategy.

The success of technological forecasting can be seen in terms of the contribution it makes to the effectiveness of that system and through it to the corporate technology strategy.

5.3.2 The approach to technological forecasting

As with any investigation of worth, the first step is to define the objectives of the technological forecast. This is followed by what is to be forecast and over what time horizon. This is basic for the study of trends which the business regards as relevant. There is also the question who should carry out the forecasting task There are various alternatives:

- a corporate staff group which is responsible for the function,
- in-house task forces established for the purpose,
- forecasting institutes or consultants.

5.4 THE MAIN TECHNIQUES OF TECHNOLOGY FORECASTING

A number of techniques have been developed for predicting what kind of developments are most likely or else for predicting how long it will be before a particular innovation triggers a change in practice. This section describes some of these.

5.4.1 Single-trend extrapolation

Specific technological parameters, such as the speed of aircraft or the thrust of jet engines are plotted over an extended period. Assumptions are then made about the expected continuation of the established trend. An illuminating example is the ever-continuing miniaturization of silicon chips for computer and semiconductor applications.

5.4.2 Substitution analysis

The investigator is concerned with the rate of substitution of one technology by another. Current examples in the car industry are the replacement of steel body parts by reinforced polymers or the use of adhesives instead of welded assemblies. For instance, 'S-curve analysis' examines, on a time base, the rate of growth in functional capability of a product or process. Its continuing growth is forecast and the extrapolation is plotted. When the comparable curve for a new technology is superimposed, a likely cross-over period is established. This marks the time at which a change in practice is likely to take off as the new technology begins a competitive advantage.

5.4.3 Scenario development

This is a description of likely actions and events leading from a set of given conditions towards a specified goal. A target scenario is developed, together with a probability estimate of its realization. Perhaps more important is the development of multiple scenarios, each with its own assumptions and likelihood level. The possible impact of each scenario on the company is then evaluated. This allows a systematic treatment of expected and less likely technological changes, which could, if they came about, have a major effect on the company.

Scenario development can include:

- global forecasting
- quantitative forecasting techniques
- qualitative forecasting techniques
- computer modelling
- environmental scanning.

5.4.4 The Delphi method

This forecasting technique uses the judgements of a broad range of experts with experience in relevant fields and chosen to maximize the cross-fertilization of ideas. A set of key questions is formulated and circulated to them. The responses become the basis for a second questionnaire which is issued with the collated answers to the first round. The second response set, in turn, provides the basis for the next round. The process is repeated until a consensus of opinion is obtained on the key questions. Anonymity is preserved throughout the exercise. The experts who differ radically from the rest may be asked to state the reasons for their views. The Delphi method is basically a survey and feedback without face to face interaction and provides informed judgement on a collective basis. Linstone and Turoff (1975) give a good account of the technique and its applications.

5.4.5 Cross-impact analysis

This approach assesses the interdependencies of technological developments. It also examines the interactions between such developments and possible social changes. Consider for instance, the development of personal computer networks and the social trend of more one-parent families. A computer-skilled mother can look after a child and work from home. What are the implications? The purpose of the approach is to generate and to assess the maximum number of possibilities.

5.4.6 Morphological evaluation

Morphology is the science of form and structure. It is used to search for new technological opportunities with a systematic, step-by-step exploration of all possible combinations of new and future technologies and applications. The evaluation can be most convenient in tabular form where technologies are listed against applications in the form of a discovery matrix.

'Morphological analysis' focuses on selected, possible developments. After describing all the characteristics of each development, every possible means of achieving the listed developments is postulated. Prediction is aided by assessing which of these means is likely, also when and how they could combine to produce the development.

5.4.7 The structural approach

With this group of techniques, structure is visualized in a tree form. On the vertical scale it provides an hierarchy of objectives, ranging from an overall social goal, through possible solutions, functions to be performed and to research programmes to provide solutions to functional needs. At each level all possible solution areas are listed. These become more specific with the descent from one level to the next. Such a structure is known as a **prospective tree** and represents a general, qualitative judgement.

A judgement of priorities can be introduced by allocating values to specific functions and criteria. Such quantification is known as a **relevance tree**. A method of priority rating was introduced by the US Honeywell Corporation, as reported by Esch (1968) This was known as PATTERN (planning assistance through technical evaluations of relevance numbers). For each level of the tree a small, mutually independent set of criteria was established, the relative importance of which was measured in relevance numbers. In numerical terms these were the product of the significance of issue 'a' to criterion 'b'. A criterion was an aspect of importance to the company, such as the strength of competition. The degree of importance was expressed by weighting this criterion. An issue is an area of application or solution the importance of which is similarly weighted by a significance number. Taking the tree as a whole, the criteria weighting and the significance numbers were expressed, so that the sum of weights (vertical) and the sum of significance numbers (horizontal) equalled unity. The statement of values requires panels of experts, appropriate for each level.

5.4.8 The limitations of forecasting accuracy

Any long-term projection includes the hazards of numerical predictions and the constraints of the economics of data collection. However, with technological forecasting there are additional complications such as:

- the interactions between technologies, especially where they are generic,
- sudden demands generated by society, such as a cure for AIDS,
- fundamental scientific discoveries, such as antibiotics, lasers etc., and the acceleration of change which can lead to technological discontinuities that invalidate the extrapolation of existing trends,
- the difficulties of estimating the rate of diffusion of a new technology and the market acceptance of the products which incorporate it.

5.5 THE TECHNOLOGICAL FORECASTING SYSTEM

The intermittent forecast, without review or updating, is unlikely to be sufficient for areas of fast-moving technologies. A limited annual review would provide continuity and can be especially useful where sudden developments in seemingly unrelated technologies can impinge on the company. It can also serve as an audit of the forecast and its underlying assumptions. Within a period of turbulence, such as with the 1973 oil crisis, the sudden change of scenarios can call for complete forecast reviews. In the absence of such a crisis a major technological forecast every five years could serve as an input to the review of corporate strategy.

There is an element of interaction between technology forecasting and the actions that might result from it. This applies especially to a 3–5 year horizon. For such a time span more specific forecasts are needed as the basis for technological improvement programmes. Forecasters need to:

- Identify not only current, but also prospective threats. Competitors are also likely to have technological improvement programmes.
- Appraise the risks, costs and problems with various combinations of equipment.

• Evaluate the implications for finance, marketing and organizational arrangements, likely to be needed to harness an established technology potential.

5.6 THE YIELD OF TECHNOLOGICAL FORECASTING

Forecasting is the precursor for planning and action. Technological forecasting provides a company with a picture of possible innovations in its fields of interest. A normative approach, which concentrates on opportunities for R&D is a framework for specifying the technological objectives which a company wishes to integrate within its overall strategy.

There is always the risk that a forecast may be mistaken, although learning experience and a systematic analysis reduce the risk. But the very exercise gives a deeper and fuller insight into the nature and the relationships of influencing factors and the sensitivity of outcomes to their variation. There are few major managerial decisions which completely exclude an estimate about the future. Such estimates can be arrived at intuitively or explicitly. It is a genius who never needs to do some homework.

Care has to be taken to assess the credibility of assumptions, the chosen horizon, the tools of analysis and the data available.

The assumptions, particularly, need to be quite explicit. There can also be the temptation that where data are sparse the available material gets a greater weighting than might be appropriate. The actual forecast can enumerate future technologies and the further development of existing ones. It can link these forecasts with probability statements. Alternatively, a time range can be given for an expected event. It would be a bold forecaster who pronounces on the future in a deterministic form and who endeavours to measure what cannot be measured.

Technological forecasting is an aid to managerial judgement. As with all techniques, the efforts and staff time expended have to be related to the resulting benefits.

5.7 SUMMARY

The inputs for technology strategy are technology assessment and technological forecasting. Technology assessment audits the technologies currently used by the company and compares these with the best industrial practice. It defines strategic technical areas, critical to the prospect of the company, and evaluates the impact of new technologies on its operations.

Technological forecasting is more concerned with the longer term trends and opportunities of scientific and technological development. One method extrapolates from existing knowledge and current established practice. The other takes a future goal and works backward to the current state. A number of techniques assess the likelihood, potential and interaction effects of specific developments. They are used to identify opportunities, threats and problems of realization.

FURTHER READING

Cetron, M.J. and Bartocha, B. (1972) *The Methodology of Technology Assessment*, Gordon & Breach, Science Publishers, New York.

Jantsch, E. (1967) *Technological Forecasting in Perspective*, OECD, Paris.

Linstone, H.A. and Turoff, M. (eds) (1975) *The Delphi Method – Techniques and Applications*, Addison-Wesley Publishing Company, Reading, MA.

Martino, J.P. (1983) *Technological Forecasting for Decision Making*, (2nd edn) Elsevier Science Publishing Co. Inc, New York.

Verschuur, J.J. (1984) *Technologies and Markets*, Peter Peregrinus Ltd, London.

CHAPTER 6

The realization of new technology

6.1 INTRODUCTION

The implementation of a technology strategy is expressed by the development and application of a new technology by the company. It must be visualized in terms of scientific principles, potential applications and artefacts, as well as structured know-how. New technology refers here particularly to product technology and development in competitive markets but it also includes the development of manufacturing processes, where this is needed for the products the firm wishes to market. Furthermore, there is no clear boundary between the development of new products and the technologies embodied in them.

The realization of a new technology, with applications of economic value, is a process of creation. This process has the following key functional constituents:

- research (R)
- development (D)
- design (D)
- a manufacturing input (M)
- a marketing input (M).

The first two constituents, research and development (R&D) have generally been coupled. The alternative combination of DDMM is equally important. All constituents matter – take one of these away and the new technology, at best, will be flawed; more likely it will fail.

The first part of this chapter is close to the subject of research management. This already has a substantial literature and it is not the purpose here to repeat what has already been well covered in some excellent texts. Rather it is the effect of technology on the very process of its development that is considered here.

It will be useful at this stage to clarify some of the concepts that are basic to the discussion.

6.2 THE CONCEPT OF RESEARCH AND DEVELOPMENT (R&D)

6.2.1 The meaning of research

Research is primarily a quest for knowledge. Within the field of technology it is also a search for the underlying scientific principles in the area of knowledge investigated. Furthermore, it provides techniques of investigation. Pure research, which may be experimental or theoretical, is such a quest, irrespective of any economic or military aspect. It is associated particularly with the scholarly activities of universities. Fundamental or basic research is at a deeper, more integrative level which may open further fields of research. The unravelling of the DNA chain is an example which has given rise to the whole field of genetic engineering. Applied research is an original or critical investigation with specific economic or military objectives; it is mission based and it is concerned with the establishment or confirmation of feasibility. This is the substance of most industrially-funded research which aims to generate new products or processes.

Generic research is a form of applied research which concentrates on the opening of 'knowledge gates' i.e. the knowledge gained has a wide range of applications, such as laser printing. The term strategic research is used by some companies to designate investigations that lead to further business opportunities.

6.2.2 The nature of development

The term 'development' describes the activities concerned with the introduction of specific products and/or processes up to and including the prototype or pilot plant stage. It also includes improvements and extensions based on the existing scientific and technical knowledge about materials, devices, products, plant, processes, systems and services, such as:

- changes of scale: larger or more powerful products, but it also includes miniaturization,
- application engineering. A typical example is the introduction of diesel engines in compact cars,
- testing. to search for, or to evaluate alternative products, processes or services involving a new technology or substantial improvements,
- the installation of new processes or systems before the commencement of commercial production or commercial application.

Although the emphasis here is on research and development activities, all aspects of technology within, and relevant to, the firm are considered here.

While in most cases development follows successful research, there are also instances where development problems have generated research programmes. Sometimes several iterations are required before technical success is achieved.

A special category is the development of know-how. This can be defined as the accumulated, practical skill, expertise and experience that facilitates effective operation. Developed process know-how can have considerable market value.

6.2.3 The distinction between research and development

As a term, 'research and development' (R&D) is widely used and also appears frequently in this book. The reason for this combination is historical. When firms were small the two functions were close; often the proprietor was an inventive genius and a research scientist did much of his own development work. While with some firms there is still some overlap between 'research' and 'development', the greater scale of operation and the increasing sophistication of science and technology has led to a division of labour in the technical operation is of the larger industrial companies.

The expression now has its problems. Research and development are lumped together, although in many firms they have become distinct activities, with different objectives, methods and style of working, organization, staff and culture. In some industries, such as aerospace engineering and pharmaceuticals the staff/years and costs of development can exceed the cost of research by factors of 10. Such differences require different control systems.

Where, because of regulatory requirements, there is extensive testing and certification work, such as in the pharmaceutical industry, the routine task components of development work can be treated as quasi-manufacturing activities, which are amenable to the use of established operations management techniques. The scale of operation may be appreciated when, for example, a UK pharmaceutical company spends £100M over ten years on a new drug for an expected £500M market.

6.3 RESEARCH AND DEVELOPMENT POLICY

It is fundamental that R&D policy and its programmes are integrated within the overall business strategy. The nature and level of integration has been stated succinctly by Roussell, Segal and Erickson (1991) in their development of the 'Third generation management of R&D'. Its essence is the working to a strategic, balanced portfolio of R&D projects, generated in partnership by general management and R&D management. The strategic purposes of R&D are to:

- defend, support and expand existing businesses,
- to establish new businesses,
- broaden and deepen company technological capacity.

This partnership is both strategic and operational. It involves the joint exploration and assessment of opportunities. The development of a shared, overall view is the basis for decisions of what research to carry out, when to do so and what resources to devote to it.

The strategic objective is to generate additional product value. At a more specific level, offensive R&D is intended to attack competitors, to improve market shares or open up new markets. The object of defensive R&D is to counter such strategies by rivals.

As we have seen in Chapter 4, a company can estimate the erosion of its markets in the absence of further product development and from this it can derive a time series of loss of revenue. In turn, such data are an indicator of the criticality of product research to the company.

6.3.1 The dilemma of target uncertainty

Abernathy (1978) makes the point that with major product introductions or innovations, market needs are ill-defined at the early 'fluid' stage of a new technology. Quite apart from the technical uncertainty, there is the risk of committing R&D resources in directions which may not yield the product performance advantage the company needs for effective competition. A company management may therefore defer research commitments in the face of both target and technical uncertainty.

The postponement of detailed and binding decisions may be an appropriate response at such a stage. According to Amendola and Gaffard (1988) this is important with opportunities about which more information is expected to appear or new opportunities to emerge due to the mere passage of time. This can be the case where improvements in technology are external and can be appropriated freely or at a given cost. For instance, the development of process instrumentation and control systems was strongly influenced by the rapid developments in microelectronics. Holding back maintained future technology options.

Where a company R&D commitment is seen as a new technology learning process is also becomes the generator of new technology options. This yields the flexibility of innovative choice. There are, therefore, advantges in setting broad goals in the first place, which can be made more specific as technological scenarios develop and the implications of generated options are better understood.Time-scales, costs and risks emerge and resource allocations can be increasingly focused on the most promising options.

6.3.2 The minimum R&D threshold

If a company is committed to research and development there will have to be a minimum level of investment in equipment and staff. Where R&D is of an advanced scientific nature, expensive specialist equipment may be called for. However, what can be more important is the minimum requirement of skilled and experienced scientists, engineers, technicians and craftsmen. Such an investment is often beyond many small to medium-sized companies. Even with larger corporations the acquisition and running costs of the research function mean that its resources need to be husbanded. The argument for some corporate research laboratories is based on the reasonable use of specialist equipment and its additional role as a 'skill centre' where a critical number of first-rate research staff can make their expertise available to the whole of their corporation.

6.4 THE STIMULI FOR INNOVATION

The activities of technology assessment and technological forecasting concern research staff and stimulus can be expected from the very tasks involved. But these tasks are by their nature intermittent and a company needs a standing arrangement for the gathering, assessment and utilization of ideas. A firm can be relatively passive about gathering stimuli. Nothing gets done unless the stimulus is very powerful or has attracted the attention of top management. Otherwise, the organization – as distinct from some individuals – is unaware of it. At the other end of the spectrum a company has a well developed technical scanning system and judges its staff on their performance in picking up ideas from the outside world.

The following are the more important sources of stimulus.

6.4.1 Customers

The uncomfortable component of this source are complaints, rejects and application problems. But these are also learning opportunities and may make a company aware of unexpected product deficiencies which, in turn, lead to research or development projects. By the very nature of product use a customer will occasionally apply and then evaluate a product in a context which was not anticipated by the manufacturer. Where such a product is equipment it may be serviced by the manufacturer. An observant field engineer and a receptive service manager may see opportunities which may not necessarily be obvious to a technical department. Well-established relations with major customers and tactful, listening salesmen can elicit a host of useful tips. The practice of some Japanese companies to second young research staff temporarily to a marketing department is to make them aware of and to value the stimulus of customer requirements.

Some companies have a formalized customer feedback system. Quite apart from the day-to-day contacts at various operating levels, top executives make periodic visits to major customers to see how their products and service can be improved. Such a relationship is typical, for instance, between component suppliers and the big car manufacturers. The focus is on the functions of the product and its key features.

6.4.2 Competitors

Any marketing man, worth his salt, manages to collect useful information about his main competitors. Much of this is customer based and informal trade hearsay but it also comes in a structured form from catalogues and exhibitions. Patent information can, of course, be a direct stimulus.

6.4.3 Suppliers

In terms of technology the range of suppliers is from a simple jobbing shop working entirely to the company's instructions to the sophisticated international corporation whose products embody a wealth of information. In some fields, such as advanced materials, the supporting technical sales and service functions provides a wealth of ideas and application opportunities.

6.4.4 Educational and research institutions

Research collaboration with a university introduces company staff to other problem approaches. It can also provide other spin-offs, such as short courses, easier access to laboratory, library and journal facilities. Again, many trade research organizations will publish regular bulletins, abstracts and relevant trade intelligence.

6.4.5 Government

Government departments, such as the UK Department of Trade and Industry (DTI) and its funded research centres provide stimulus and encouragement through publications and conferences. Although the stimulus is often free issue to all interested parties is is also non-proprietary and does not necessarily provide domestic competitive advantage. It is the resultant perception of opportunities which makes such stimuli valuable. Also, the role of government as a major equipment purchaser can be a significant source of stimulus. Some of these aspects are further developed in Chapter 16.

6.4.6 Internal stimuli

The very research activity within a company creates its own stimuli. New experiments provide new insights. Pilot plant findings point to new avenues for investigation; development problems or manufacturing difficulties are a common source of impetus. The 'bubbling-up' of research ideas can also come through suggestion schemes, sometimes from quality circles and the informal know-how trading between engineers and scientists at conferences.

6.5 THE SOURCES OF INNOVATION

It is often assumed that most of the product innovations are generated by the company that manufactures them and that it is a part of its intelligence system to perceive product opportunities.

Research carried by von Hippel (1990) however indicates that this is by no means so and that there are substantial variations between different industries. For instance, out of a total of 111 reported innovations relating to scientific instruments 77% of the innovations came from users of the equipment and 23% from the manufacturers. Often the users were university laboratories where there was a need to measure or to control a new process. The innovation and sometimes the prototype were developed by university researchers to meet their own requirements. Typically, a manufacturer was asked to further develop the instrument for commercial exploitation. This research implies that many companies should look carefully at their customers for innovation opportunities.

By contrast, in some fields, such as engineering plastics, the materials manufacturers are the dominant innovators. Pavitt (1984) provides an helpful classification of different UK industrial sectors which are major sources of technological innovation. He lists these as follows:

1. 'Supplier-dominated' sectors. The innovations are mostly process innovations, embodied in capital equipment or industrial intermediates. For instance, in the textile and shoe industries these include machinery and synthetic materials.
2. 'Scale-intensive' sectors. Innovations mainly come from large companies, operating with significant economies of scale, such as with food products and some consumer durables.
3. 'Specialized suppliers.' These are mainly modest specialist engineering companies, working closely with customers.
4. 'Science-based' sectors. Industries with large R&D programmes in rapidly developing fields, such as electronics or advanced materials.

6.6 THE INTELLIGENCE FUNCTION OF R&D

The research director of a major UK aerospace company with over 20 years experience in R&D came to the conclusion that 99% of research ideas came from outside his organization. Chance observations in the outside world accounted for 50% of the ideas, while the passage of information came to 49%. The former was a function of attitudes and perception while the latter was to some extent a matter of organization. The remaining 1% was arrived at by logical deduction.These percentages reflect the numbers, not the worth, of ideas but they are a reminder that no company or institution has a monopoly of inventions and innovation, even in a field in which it is a research and market leader. Many of the great opportunities for innovation are based on knowledge acquired by others. The most profitable opportunities and the most serious threats sometimes come from looking at old problems in entirely new ways – not from traditional approaches. it is therefore important for a firm to look outside itself; to gather ideas and stimuli. Furthermore, a framework of quick evaluation and, if appropriate, absorption is required.

It is important for a company to develop the intelligence and learning skills of its R&D staff, so that disciplined, structured information tracking and assimilation become accepted custom and practice. The effectiveness of learning depends on experience, reflection, conclusions and implementation. Well-chosen research teams have members who can readily contribute to these elements.

6.6.1 Technology scanning

This is a systematic approach to keep abreast of all technological developments relevant to a business, the opportunities they offer and the threats they contain. While different industries draw on different types of knowledge bases, scanning needs to be more than the intermittent, informal perusal of trade journals. Although scanning for opportunities is a common part of the entrepreneurial role of senior managers, with some companies technology scanning still tends to be an *ad hoc*, activity only undertaken when they become aware of a sudden threat or problem.

Technological scanning requires an operating system. What is done with an opportunity depends on how it is defined. There needs to be a structured view about the perception, collection and internal dissemination of technological intelligence. Not only should the scanning apply to the technologies which the firm uses but also to 'parallel' technologies where comparative problems and possible solutions may be found. A 'gatekeeper' responsible for monitoring developments is required and that role must be supported at senior level.

Normally, an experienced, well-structured R&D team has better opportunities of spotting trends and incipient threats.This applies both to small incremental developments, application opportunities and more radical changes. Intelligence presumes not only the acquisition of information but also its comprehension. Its implications must be understood and constitute the basis for action.

Nelson and Winter (1982) point to the process of incremental learning in which success indicates the use of particular search and design procedures and concentration on critical product or process features. Development along such lines leads to the perception of technological trajectories. Not only does the firm have to identify which technologies and scientific developments offer opportunities through comparative advantage but also to ensure the requisite skills are available in-house or can be assimilated from external sources.

Intelligence concerns not only external but also internal developments. Where R&D is physically remote from manufacturing operations, potentially promising ideas can be overlooked. There can also be the segmentation of knowledge between the various divisions of large corporations.

6.6.2 The role of technical information services

The provision and careful integration of library, journal and abstracting services is an important contribution to research intelligence. The ever-increasing volume of scientific and technical literature requires selection and predigestion. The range and the size of databases are continually increasing. The challenge is to capture a stimulus but still to contain the scanning costs. The publications of research and trade associations can be invaluable here. A close link between information services and research groups, coupled with careful scrutiny of published material, can reduce duplication of effort. There are no fortunes in reinventing the wheel!

6.6.3 Patents as a source of technology intelligence

We have already noted in Chapter 2 the importance of patent information as a source of technology intelligence. There is more to patents than the latest moves of competitors. In some fields of rapidly changing technology, such as electronics, some leading companies have systematic search programmes as part of general intelligence. The wider stimulus of patents in relevant fields of technology can yield suggestions for research and structure solution possibilities for technical problems. For instance, a set of patents can form the basis of a brain-storming session. In a recently reported case, a medium-sized Swedish company carried out a patent search on a new dust removal device. It found that its design ideas had already been anticipated by several patents. However, the information and stimuli provided by these patents allowed its engineers to completely rethink its dust removal problem and yielded a new device which the company patented. In cases where no such solution could be anticipated it might still be preferable to pay royalties for the use of existing patents than struggle into a misconceived research programme.

Again, the analysis of patents, especially where they form clusters filed by a particular company, gives indications of intent and possible market developments with new products in, say, two to five years.

Patents also serve as a source of intelligence when it comes to the negotiation for a licence to use a particular technology. The licensee expects the benefit of the licence to correspond to a given period of usage. It is possible that the assessment of patents filed by the licensor could indicate prospective developments which could reduce the value of the licence e.g. the replacement of an electro-mechanical device by an electronic processor. Armed with such knowledge the prospective user could well negotiate a different royalty deal.

6.6.4 Technology capture and transfer

The first task is to trace the sources of the target technology. Individuals and organizations which can provide potentially useful knowledge need to be identified. Contacts can be through formal or informal networks, such as computer clubs, local productivity associations, trade associations etc. Next, there is the tapping of these sources, such as the systematic scanning of abstracts. This is followed by the absorption, the transfer of knowledge for use within the company. It has to reach those who can, or will be trained to handle it. Of course, an important transfer opportunity is to hire staff with the appropriate expertise.

Effective and continuous technology capture requires a suitable company climate which encourages curiosity and interest in diverse technological areas. Some individuals have a natural tendency for this; they will get to know the latest developments as well as where these are taking place. They provide both knowledge and contacts and have established a role of gatekeeper.

6.7 THE MANAGEMENT OF R&D

The management of R&D can be divided into two parts. The first is the organization of the ongoing functional activities. This will be described here. The other is concerned with the management and organization of specific projects and will be discussed in Chapter 8.

One of the tasks of management is to ensure that operations express and implement company technology strategy and policies. This requires some care. The interpretation and filling out of a strategy decided by senior management is coloured by the discretion exercised by intermediate management levels. Their decisions can, intentionally or otherwise, reinforce or undermine company strategy. The degree of staff participation at the work place level is often a function of middle management attitudes. Much depends here on how a new technology is developed.

6.7.1 The role of top management

Top management here has two major responsibilities:

1. The overall direction and control of the R&D effort,
2. the selection of key managers to be entrusted with major programmes.

The implementation of technology strategy requires a corresponding direction and surveillance at company board level.The first responsibility is illustrated by Malpas (1987) who described the organizational framework for such a role. His company, British Petroleum International plc, saw the size and scope of the company research and engineering effort as a major issue. R&D was overseen by a research board which had the same status as the business boards. It comprised the chief executives of the major businesses within the group, the heads of corporate planning, R&D and of engineering. There were also two non-executive directors from the main BP board, both distinguished scientists and industrialists. It was the responsibility of the R&D board to determine the appropriateness of the total R&D and where it was to be carried out, within or outside the company's laboratories. It took into account the size of each constituent business and considered its short-, medium- and long-term relevance.

About 15% of BP's total R&D budget was at the corporate level, much of which was devoted to projects which cut across business boundaries. A special directing group, which included the chief executives of the businesses concerned monitored these with particular reference to longer term objectives and funding requirements.

The second major responsibility of top management is emphasized in those industries where product life cycles are short, such as in computing and the electronics industries. Here, programme managers have a crucial role. They need to be steered and supported by top management in the early stages of the product cycle. After that top managers have to stand back and not get involved in details.

6.7.2 Organization structure

Chandler (1962) stresses that organization structure follows strategy. It is a means of implementation. Without a corresponding organization structure the strategy is at risk. The structure of an organization can be defined in terms of the tasks and functions of each of the constituent units and their relationships with each other. Organization structures affect innovativeness. Whatever the context, it is important to have the innovators as close as possible to the commercial processes. Where a company strategy is 'technology push' then there must be close integration with the R&D function. If the strategy is based on a 'demand pull' approach to technology close relationships with marketing are essential.

The main functional departments of a company that are usually involved with the development of the new technology are: research, development, design and manufacturing/process engineering. Their organizational description and position vary with company, industry and time, but in sum they constitute the substance of technology management. Some sample organization structures are shown in Figs 6.1 and 6.2. For illustration the charts show functions, typical of the process industries as well as for the engineering industries.

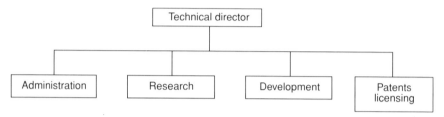

Fig. 6.1 Structure for technology management.

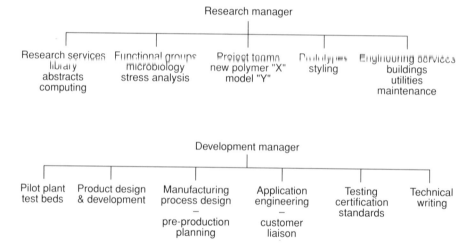

Fig. 6.2 Sample structures for research and development.

Where the R&D establishment is on a major factory site the link with manufacturing engineering is often much closer. In many Japanese companies R&D is team-based rather than functionally organized, and forms an integral part of the product development system, involving engineers from all the important functions in the firm.

The relationship between technological change and the organization structure of industrial companies has been well developed since the seminal work by Woodward (1965). Of particular interest here are the organizational implications for the R&D function itself. An example of this was the reorganization in 1987 of the Volvo car development structure. The function had to contend with the growing size and complexity of the total company organization, the greater technological complexity of products, e.g. computer-controlled engine management systems, the need to develop the quality and flexibility of the development process, quite apart from lead time and cost reduction. There was also the impact of development automation, such as computer-aided design (CAD) with its great potential of streamlining and shortening the design process. Furthermore, the ever-present need to stay competitive in the market place increased the number of product lines and its variants.

One form of organization which is often used within the R&D function is an organic management structure. The structure is task-oriented with a low level of specialization. The division of work is variable and there are few levels of authority; the organizational pyramid is flat and the span of control is wide. In the work context there is little formal specification of duties, rules and procedures. Authority is decentralized and is knowledge, rather than position based. Interaction between different units at operating level is strong and informal relationships are emphasized.

An important role within an informal organization is the 'intrapreneur' who is the product/technology champion. He may have started as a researcher but has the qualities to manage and promote all the other tasks that are needed to make the technology or the product incorporating it a success.

A significant distinction between research and development activities has also to be taken into account. Research work needs to be inspirational and is knowledge-oriented. Much of development is relatively routine work and is product focused. They need different organizational frameworks.

6.7.3 The organization of R&D – the multinational corporation

The strategic organization and location of R&D becomes particularly important for the international corporation. By virtue of their size, composition and geographical spread, serving many different markets, the multinationals have more opportunities, but also more risks. For them the degree of autonomy accorded to divisions and subsidiaries is a major facet of business policy. A new ventures department (NVD) has been used by some major corporations as a central unit to identify, sponsor and commercialize new products and to acquire new technologies. An example of this is the overall research organization of Siemens AG.

Within that company each business unit has its own market-oriented R&D. In addition, the company founded in 1991 an 'innovation centre' within its corporate research and development function. The purpose of this centre is to develop new business alternatives and to identify the technological base required for the next 5–15 years. The first two innovation projects of the centre are universal personal communication and mobile energy supply.

6.7.4 The style of management

A number of writers have emphasized the requirements of a suitable management style for research and development. A participative and flexible form of leadership is seen as the most appropriate. Task and role are more important than rank and title. One successful UK manufacturing company emphasizes leadership, not management. Particularly at shop-floor level, the connotation of leadership is more positive than that of supervision. Formal structure and hierarchy is kept to a minimum. There are few levels of authority. Formality is low. Motivation is by involvement and the satisfaction of psychological needs. The initiative and creativity of subordinates is encouraged. Instructions are given in task and objectives terms, supported by advice and information. Areas of agreement are stressed, differences are kept in a low key. The emphasis is on relaxed, businesslike cooperation between specialists who stimulate each other, communicate their problems and often collectively seek their solution. The informal transfers of know-how and the learning opportunities these provide accelerate the technical advance of the company.

Nevertheless, as Twiss and Goodridge (1989) stress, managers have to be clear about the basic goals of the company; to create and to nurture a climate or culture in which the creative can create; control the controllable; separate the maintainers from the developers.

6.7.5 The management of innovation

The larger the organization, the greater are the challenges of bureaucracy. A framework of established rules, however essential to the conduct of a business, can often have an inhibiting effect on the initiatives of staff. There also has to be a framework to encourage innovation. Often it is a perceived need and the solution to this need which leads to innovation. It is a management responsibility to provide the avenues and motivation for this to happen.

Furthermore, the complexity of innovation activities poses a challenge to the R&D manager who needs to focus his attention on a limited number of key variables. The span of human judgement is limited in terms of effectiveness to a small number, say 6 to 10, unidimensional judgements.

The proper management of innovation has an important implication. Innovation cannot be pursued obsessively at the expense of other competitive considerations, such as the timing of product launches. If too many innovations are adopted together, product costs may exceed budgets and, at least in the short run, the company may have the unpalatable choice between insufficient margins or a price level which curtails demand.

6.7.6 Management learning

It is an essential complement to technology strategy that managers and staff at all levels are able and motivated to continuously acquire new skills, be they related to the management of technological complexity or relatively routine operations. As markets and technologies change the adjustments within the organization need to be natural.

Managers have to learn at the tactical level, such as the solving of problems with a short time-scale, as well as at the strategic level, where a major competence will be to marshall opportunities and to transcribe them into effective projects. This requires realistic goals, well chosen key staff who can operate within a well-designed decision framework, set targets and limits.

An adaptive learning outlook is important to cope with technological uncertainty. Anticipating the imminence of technological change is a key managerial talent that cuts across the functional boundaries of the firm. Firms have to continuously scan their environment to be aware of and to evaluate potential technological developments that impinge on its operations. It needs to manage an ongoing portfolio of opportunities and threats.

6.7.7 The organizational climate

Structure and style of management determine the 'climate' of an organization. Calori and Noel (1986) note that companies in emerging high technology industries seem to prefer an organizational style with the following key features: team spirit, personal commitment, mutual trust and rewards for success.

Most scientists and engineers like a stimulating environment to harness and apply their creative capacity. The greater the uncertainty with a new technology, the more organic needs to be the organizational design. The less codified the technology, the greater is the influence of those few who are experts. If the new technology is still difficult to define in terms of scientific laws, rules or formulae, the more important will be individual know-how. The choice and management of staff will be critical as the company may become hostage to them.

On the other hand, with complex technology, where a number of scientific disciplines are involved, the role of the technology integrator becomes focal. It would also be his role to harness the synergy that is potential between different disciplines. This is derived not only from the various bodies of knowledge but also from the interaction and experiences of its various practitioners.

6.7.8 Requisites for success

The successful execution of R&D depends on a number of factors. To start with, there need to be effective communications within the firm, especially with manufacture and marketing. Close links with key customers can signal problems or promise. Much depends also on the personal characteristics of the innovation champion, his experience related to the main business functions and his standing within the organization.

The following are more specific requirements for success:

- full support by top management,
- a full-time leader,
- careful team building,
- common team location,
- careful team briefing and training, where needed,
- challenging and clear targets,
- focus on monthly reviews with senior management,
- a system of speedy and flexible decision making.

6.8 THE R&D TEAM

The essence of a team is the shared commitment and ownership of a clear purpose. The justification for a team is its synergy. It has to achieve more than its members would if they worked as separate individuals. One of the important technical management skills is effective team development. This builds, wherever possible, on what people are already doing, uses organizational strength to advantage and, within an overall framework of intent, encourages the self-setting of objectives/milestones. The team should not be too large. For one research director a group of 3/4 research workers, drawn from different disciplines, was the most effective while in other companies groups of up to ten research staff were regarded as appropriate. Often, the actual numbers within a team will vary with the state of the project. Some specialists may become team members on a time-share basis with other projects or professional duties.

Technological changes have made research problems and projects more interdisciplinary by nature. There is therefore a need for technologists with a fundamental grasp of scientific and engineering principles but with a broad outlook. While a blend of different disciplines is an asset to a research group, there is also some scope for different personality profiles. The forward-looking, risk-taking creative thinker can be matched with the cautious analytical mind and the risk analysing, feet-on-the-ground decision taker. To maximize interpersonal stimulus and reduce communication problems it is best for team members to work alongside each other.

6.8.1 Team management

Effective team management has to reconcile two basic requirements in job design. The first is to retain the freedom of scientific and professional staff to exercise their particular expertise and judgement. Often there is a need for innovators to form closely-knit groups with substantial autonomy, yet the group and its requirements must not be isolated from the mainstream of the organization. The second is the effective alignment of specialist groups to the research programme and the work of other groups.

There is the internal team management, with its portfolios of projects, work packages and staff assignments. Furthermore, there is the management of relations between the team and other teams and functions within the organization.

(a) Leadership

The choice of the team leader and key members is often so crucial that it may be better to accept limited delays until the appropriate persons become available rather than to proceed with less suitable staff.

Where there is spontaneous leadership within a group it develops a problem-solving approach – the group acts as a problem-solving coalition. The composition of the group also evolves relationships with its surroundings. With established channels and methods of communication the group will be stable as long as its attention is focused on its main problem.

(b) The harnessing of good ideas

To maximize stimulus and ideas generation it is important for a research team to have a mechanism by which insights obtained in one discipline are effectively transmitted to practitioners in other disciplines. Although a research team may focus on a specific project it is not unusual that its investigations lead to other ideas and opportunities. Not being relevant to the project these could be neglected. To avoid such opportunity losses one major British company gives its researchers a 'hunting licence' to capture product ideas, wherever they came from, that could be used within the group. The hurdle for an idea was £10M in expected sales.

6.8.2 Aspects of motivation

For a few individuals the level of inspiration and focus on achievement transcends everything. There is little need to motivate such a person. For the performance of the majority, motivation matters and, other things equal, a well-motivated group of employees is likely to achieve more in shorter time. Motivation, therefore, is a contributor to technological progress within a firm. Awareness of Maslow's 'hierarchy of human wants' (1943) is as relevant to an effective manager in R&D as it is for any other management role. The hierarchy, from the basic physiological needs and the requirement for security to the more complex wants of a sense of belonging, self-esteem and self-actualization seem basic to every individual. The spectrum of wants of the gifted and ambitious young research worker is different to that of the technician carrying out routine, but important tests. Overall, the culture of R&D is different to that of manufacturing and the observant manager will appreciate that different cultures can attach different meaning and value to each of these classes of need. Conflicts between individuals and groups may therefore be generated by the different perceptions of these needs.

The need to obtain psychological satisfaction from work within technical departments has not had the same level of attention as it has in manufacturing plants. In any case, there are usually more opportunities for job rotation and job enrichment than on a mass production assembly line. Nevertheless, Herzberg's 'hygiene-motivation theory' (1959) is pertinent. Motivation needs a satisfactory work environment. This includes conditions of employment, status, company policy, salaries and the physical surroundings of the work place. Herzberg calls these conditions 'hygiene factors'. These have to be suitable, but by themselves are not enough. In addition, there is the need to satisfy the inner drives of staff which include the quest for real achievement, work satisfaction, recognition for good work, responsibility and the opportunity for personal and professional growth.

In some fields, such as computer software and electronics, the early stages of a new technology can be strongly people-dependent. The key workers may be tempted to leave and set up on their own if the entry barriers to the new technology are modest. There is therefore a particular problem of managing and motivating such staff to stay.

6.9 THE EFFECTIVENESS OF R&D

With the escalating costs of R&D and the inherent risks in a competitive product market, the effective use of research and technical staff is a basic requirement of technical management.. A good manager will remember that efficiency is doing things right and effectiveness is doing the right things. Both are components of R&D success. He will also be aware of the distinction between the longer term problems and the 'quick-hit' opportunities where an intensive concentration of staff for a relatively short period can resolve more pressing problems or secure an important time advantage.

Particular attention needs to be given to the deployment of key staff. For instance, there is a distinction between core designers who develop the key product design and other designers whose work is more of a routine nature. The core designers embody the company's accumulated knowledge and design experience. They have ability and standing within the firm and are generally difficult to acquire. Their use needs to be husbanded.

6.9.1 The assessment of R&D effectiveness

Success in R&D can be assessed in technical, management, commercial and economic terms. Technical success is attained when the technical objectives of the project are achieved and the task specifications have been met. Management success is achieved where this has been accomplished within budget and programme schedule. Commercial success is determined by the performance of the resultant products in the market place, by the extra sales achieved and the contributions these make, or by proven processes in use. One measure of R&D effectiveness with ongoing operations are the annual savings in operating costs obtainable without appreciable capital expenditure. Where such expenditure is needed then it would be judged by the return on the ensuing investment. Economic success is measured by the rates of return on the R&D expenditure once a product or process is launched. Technical success is a condition, but not a sufficient condition, of R&D success. Where a company undertakes a number of research projects, its R&D success is judged by the performance of its total portfolio, such as the number of patents applied for and granted and the royalties received.

One example of research success was the experience in the 1960s of a well-known UK chemicals manufacturer. It successfully developed a process which yielded a licensing income that more than paid for all its R&D expenditures. In that sense, the process became 'free issue' to the operating divisions, quite apart from the inherent process gains of the new method.

A five nation survey undertaken in 1989 by PA Consulting Group reported on the criteria of judging the effectiveness of R&D (PA Consulting Group, 1989). Respondents from the United Kingdom and the then West Germany put the revenue contribution from new products first while France and The Netherlands put forward the number of new products as the first criterion. Japanese respondents considered return on investment as the most important factor.

6.9.2 The productivity of R&D

The continuous advance of technologies has also increased the productivity of the technology-creation process. For instance, the wide availability of powerful computers and software packages, such as for computer-aided design has increased the productivity of R&D and design operations, once effective learning and training has taken place. There are also great opportunities for increased productivity and cost reduction in the automation of routine tasks within the development function, such as the use of robots for round-the-clock testing of drugs.

Over the last 10–15 years one large German electrical engineering group has achieved R&D efficiency improvements of 6–14%. Furthermore, it expects to achieve rationalization gains in R&D which are comparative to the rationalization of manufacture. In some cases, the group has reduced the lead time and costs of R&D by a factor of 2 with the use of simulation instead of building and experimenting with equipment prototypes. In one instance, the difference between the simulated pre-production unit and the actual performance of a manufactured electronic filter was no more than 0.1%.

6.10 MARKETING ASPECTS OF R&D

It is fundamental that the research and development specification for a new product pays attention to the marketing aspects of the product functions and attributes. There can be further market-related points. For instance, the schedule of task completion has to take into account the timing of market entry, such as an important exhibition. With technological alliances and the operation of multinationals there often has to be international coordination of R&D projects. One development aspect which can yield commercial opportunities are 'product tack-ons' such as packing, printing and point-of-sale advertising, such as patient advice with pharmaceuticals.

The *Financial Times* of 2 July 1992 reported that Japanese electronics companies were forced by the downturn of international markets to reconsider the role of market research in focusing R&D. Product life cycles had become too short and product ranges were too wide.

Large annual percentage increases in R&D spending coincided with the surge of the Tokyo stock market where funds could readily be raised. With the heavy decline of this market in the first part of 1992 and the turbulence within the Japanese financial system, pressure increased on these companies as the cost of R&D finance increased.

Companies no longer wished to spend too much development money on the minor upgrading of existing models in already overcrowded markets. A shift of control over R&D funding is taking place from individual 'profit centres' to administrative divisions which have broader access to market research. Group business planning divisions prepare market studies which are circulated to R&D staff. To emphasize the close links between R&D and marketing, NEC, the Japanese electronics company, launched in 1991 its 'customer satisfaction concept' as part of its R&D policies. All members of R&D staff are expected to have a 'marketing mind'. Linked to this approach is the concept of market driven quality (MDQ) and the view of a product defect as 'failure to achieve customer satisfaction'.

An interesting variation of this policy is reported by Funk (1993). Some Japanese manufacturers install prototype equipment models in customer factories to obtain feedback from an actual operating environment. For instance, design engineers for packaging equipment work with customer process engineers and use the experience for the immediate updating of their equipment and systems. Initial performance on yield and downtime is gradually tightened with the progress of work after which an equipment specification is agreed.

6.11 FINANCE FOR R&D

This section is concerned with the allocation of funds to implement corporate technology policy. The provision of funds and the budgeting for ongoing R&D is a matter of concern for many companies. A number of different policies are used in practice.

The first is the allocation of a percentage of company revenue to the support of R&D. In its simplest form this is a mechanistic rule without a strategic dimension. Sometimes the record of R&D spending by a company may approximate such a pattern over the long term but there is no specific argument why it should be so. Such a policy is more often a measure of affordability than opportunity or strategy.

A second policy is to devote a percentage of profit to R&D. Apart from the argument that in better times a firm is able to afford more outlay on R&D there is little to be said for this. The vagaries of business results are a poor basis for a R&D programme, quite apart from the accounting aspects which can enhance or reduce profits for quite extraneous reasons.

The third and most relevant approach is to implement the overall business strategy of which technology strategy and R&D policy are integral components. The total required finance will be the sum of authorized projects and other defined missions. Of course, as with any other functional activity, the R&D budget is a part of the total financial plan, both for the given financial year and a specified horizon.

6.11.1 The use of R&D funds

A telling account of how R&D funds were spent came from a former research director of a major UK aerospace company. In his group the cost pattern of engineering R&D was broadly as follows:

1% basic research
9% applied research
90% development.

While this cost apportionment varies with technology and industry, it is the development function that generally takes the major share of the budget. This is supported by a survey by IMR for KPMG Peat Marwick High Technology Practice in 1992, based on interviews with 324 high technology companies that carry out R&D in the UK (IMR, 1992). It reported that 105 self-described technology leaders spent 68.4% of their research budget on product development and 6.3% on pure research. For 65 self-designated technology followers the figures were 77.0% and 3.6% respectively.

In view of the ever-increasing costs of R&D the containment and control of development expenditures is a major management task.

6.12 DDMM – DESIGN, DEVELOPMENT, MANUFACTURE AND MARKETING

An alternative approach to the research and development concept, particularly relevant to the engineering and allied industries, is the emphasis on and the integration of design, development, manufacture and marketing (DDMM). The combination of these constituents has been termed 'downstream closeness' by Ansoff and Stewart (1967). This is governed by the extent to which the success of the company's product introduction process depends on the communication and cooperation between the R&D, the manufacturing and the marketing functions. In many technology-based industries such a strong coupling is imperative. For instance, with the manufacture of advanced polymers, where much depends on application engineering, the functions of product and process development work closely with production and field technical service. Effective information flow between all these functions is vital.

While there often is great emphasis on the close association between design, development, manufacture and marketing, such an integration can be further fostered by giving the task of shepherding a project and ensuring integration to a senior executive who will be the product/project manager and champion of the new product.

6.12.1 The role of design

Product design is concerned with the creation and shaping of a product to a functional specification, which incorporates competitive value and manufacturing feasibility. Process design, including tool design, applies to the tangible realization of a process specification. Neither needs to incorporate new technology. For instance, the design of a car for a left-hand instead of a right-hand drive is the permutation of existing knowledge and techniques. Much design is the matching of existing technologies to a new context or use, such as the design of a car for specific arctic conditions. Of course, there are many design projects, such as a new aircraft, where major technological advances are the very essence of the venture. Design, particularly engineering design, is a key component of the development of new technology.

The classical view of R&D gives insufficient attention to the role and contribution of design. Often it is subsumed within the concept of development to which it is closely related. Design appertains to products, systems and processes. It applies more to artefacts than non-dimensional materials, such as chemicals. Design activities range from sophisticated intellectual activities, using advanced mathematical and computing techniques, to relatively routine drafting tasks. Design automation of repetitive tasks has transformed the operation of design offices; computer screens and networks have replaced the drawing board. Software packages and expert systems have simplified many design tasks. Increasingly, the design function has become integrated with other technical and manufacturing functions.

The actuator of the design process is the project specification; the requirements of senior management of what is to be done, to what budget and by what date. It is a commitment to the totality of a design. The formulation of such requirements is often the result of previous R&D activities; the basic technical problems are presumed resolved. The distinction made here is that design is based on the assumption of feasibility whereas the task of R&D is to establish such feasibility in the first place. The distinction in principle is somewhat fuzzy in practice, but broadly speaking it holds. For instance, feasibility analysis often involves preliminary design studies.

The management of design is a function of technical management. Design work is commonly organized in project form with budgets and task schedules. Functional specialization is common in large design offices. Design is particularly concerned with detailed product development for manufacturing feasibility on the basis of established process capabilities.

This section concentrates on those aspects of design which are particularly relevant to the management of technology. A list of appropriate texts is given at the end of this chapter for readers who wish to study the subject of design more fully.

6.12.2 Some features of the design process

Design was once described as decision making with a pencil; now it is with a mouse or a light pen on a screen. While the role of intuitive flair is acknowledged, the management of design is best appreciated as an hierarchical decision process. This goes from the totality of an overall concept to a multitude of detail design problems and trade-offs. The following constituents are relevant to the development of new technology.

(a) Conceptual design

This is closely linked with the product specification the feasibility of which it has to confirm. Its main activities include the following:

- the definition of the major design parameters, such as key functional properties or dimensions;
- the sensitivity analysis of these parameters, such as the effect of car engine size on the design of car exhaust systems;
- the compatibility of subsystems, such as the relationships between a car engine, transmission or suspension;
- the allocation of factor values for the purpose of trade-off and optimization studies.

This stage of design is concerned with the main product functions and attributes and has to anticipate customer expectations. These can only be realized by close collaboration with corporate customers and detailed market research for consumer goods. Here is the scope for creativity, to incorporate the latest relevant technological developments. A number of creative techniques have become well established, such as brain storming, synectics, idea diagrams and morphological analysis. These techniques help to establish options at the fluid stage of design development. It is here where regular technology scanning yields candidates for evaluation.

(b) The use of modular design

The consistent use of modular design maintains simplicity of, and reduces uncertainty in, manufacture but can, nevertheless, yield a large finished product variety. It cuts down the design and test task content of a product development programme and permits the carrying forward of current component families to the next product model.

(c) The design review

This is important for the early proving of a design and the detection of design weaknesses. Such a review applies to the component, systems and overall design levels. The pressure for quicker product launches emphasizes rapid prototyping, testing or simulation to confirm design calculations and conformance to quality and safety standards. At this stage, there needs to be close collaboration between design and development departments. Just as in manufacturing assembly there are error avoidance techniques so design error prevention routines are important. Where these are included in computer-aided design (CAD) software packages they can accelerate design confirmation.

(d) Design within an activity accounting system

Some of the tangible benefits expected from an activity accounting system (see Chapter 10) is that designers will use activity cost driver rates for a fuller understanding of the cost of their design. The design engineer 'owns' his cost; close links with manufacturing engineering will reduce these. With CAD systems, process times and manufacturing cost rates can provide a quick design decision guide.

(e) The integration of design and manufacture

The interface between design and manufacture often requires special attention. For instance, at one major UK electronics factory it was the responsibility of the design authority to take into account manufacturing facilities and capabilities. This was assisted by a product introduction group set up within the manufacturing facility who would, during the design and development stage, work mainly on the design authority site to ensure that manufacturing requirements were met. When the product went into pre-production build, key staff from the design authority would work within manufacturing to ensure that engineering changes were progressed as quickly as possible and to secure that product to market targets were kept.

In some cases, such as with a new and incompletely developed product the quantities involved, say, for test marketing, are often so small that production is carried out by the R&D team along with development of know-how.

(f) The use of knowledge-based systems (KBS)

It is the experience of some companies that less than 20% of design time actually adds value to a design. Designers spend most their time searching for and reformatting information. The development of comprehensive databases, such as engineering data management systems (EDM) and computer-aided design (CAD) has significantly improved design productivity by providing a readily accessible and integrated flow of information for a design team.

They have also enhanced the opportunities for integration within DDMM systems. For example, CAD/CAM provides immediate transcription facilities for the detail component design into tool drawings, computer-controlled machine instructions, quality assurance and automatic test systems. Furthermore, automatic feedback from manufacture and quality assurance allows the prompt revision of detailed design and the generation of all the required documentation for engineering changes. For complex assembly operations, such as with aero engines, digital pre-assembly processes facilitate concurrent assembly mock-ups where designers can see instantly the effects of a new or relocated component. Upon confirmation these can be used for the actual assembly operations. Moreover, data sharing over local and remote networks makes key information readily available to other company establishments. All this has substantially reduced overall and detailed design times. A well developed database now constitutes a major company intellectual property asset.

6.13 THE REDUCTION OF DEVELOPMENT LEAD TIME

There is the temptation for research staff to aim for the 'ideal' solution or to carry out parallel investigations which exploit the overall potential of an invention. These endeavours can lead to heavy expenditures without corresponding benefits. Again, engineering development is a process of successive approximations. In some cases the time required for a 'perfect' decision is not justified when compared with the time scale of a good solution. It is therefore crucial for management to insist on commercial objectives and specifications which will achieve these.

In some very fast moving markets, such as consumer electronics, the need for early product launch tends to be the dominant requirement. Where the product life cycle is short a six-months deferment can cut a substantial slice out of the life cycle profits of that product. A high concentration of resources may be needed to achieve a market entry before competitors. The temporary market leadership which such a time advantage yields justifies a degree of development cost overrun.

Competitive success depends more and more on time-based competitive capability. As an instrument of global competition 'the time to market' covers the total product development process from the initial product concept to the establishment of the planned product market share. It is based on established product and technology strategies, is customer driven and is characterized by a collaborative working style where product designers meet customers and discuss their problems.

The closeness to key suppliers and customers is illustrated in civil aircraft development.British Airways engineers, seconded to the Boeing aircraft company as customer technical representatives, participate on the spot in the detail design of the Boeing 777 aircraft. With customer-driven product development the customer is involved in the design specifications and their subsequent interpretation. Operational needs, assessed through the product life cycle, get immediate attention.

Another example is the policy of some Japanese electronics companies to minimize design and innovation risks. Careful design for manufacture is coupled with the 'rugby' approach of working in well-drilled project teams. A hand-picked, multi-disciplinary team works closely together from the start to the finish of the development process. It has a high degree of autonomy and contains representatives of key suppliers. These not only include component manufacturers, but also tool makers. One of the most important features is the early release of tentative design data for long lead toolings, such as press tools. Crucial trade-off decisions are taken as early as possible in the development process to reduce the number of engineering changes close to product launch when they become much more expensive to carry out.

The final stages of product preparation are particularly important. This is when some Japanese companies, such as Toyota, apply the 'ramp-up' approach. This focuses on intensive problem solving and debugging on production start-up. Design and manufacturing engineers work closely with production control and quality management to iron out last-minute problems and make on-the-spot engineering changes if needed.

A fast system of product renewal to capture initial market opportunities is helped by flexible manufacturing technologies. Proven product improvements are batched and introduced at designated times in a period of intensive adaptation to meet marketing updates such as model upgrades. Being first in the market with a new or upgraded product can more than balance the limited experience curve effects and the ability to bring manufacturing costs down.

As the time value of money matters, another relevant benefit of lead time reduction is an earlier reversal of project cash flow, leading, other things equal, to a quicker recovery of outstanding development expenditure. The financial aspect of lead time reduction and the implications for project management are further considered in Chapter 8.

6.13.1 Concurrent engineering

This applies to all development activities from product feasibility confirmation, at the pilot project stage, to volume production. Initially applied to a single product it is now used on multiple product lines. Both the terms 'concurrent engineering' and 'simultaneous engineering' are in use. They denote DDMM systems where the development work for a new product is carried out by the relevant departments in parallel and not in series. This reduces the product development lead time. Both product and processes are validated at the point of design so as to eliminate redesign as the development passes from one phase to another. All product life cycle requirements are established at the concept definition stage.

Where concurrent engineering is based on the 'rugby team' approach, product planners, marketing experts, designers, both product and production engineers and key suppliers work together in project teams from the beginning to the end of the product development process. Responsibility is given to those who add value. Particular leadership skills are required for the management of such cross-functional teams. These must be empowered to achieve their objectives; they also must be tracked and held accountable for what they are doing.

One of the challenges of parallel working are the risks incurred when some activities, such as manufacturing planning, proceed before all aspects of detail design have been resolved. The very planning and scheduling of the concurrent engineering tasks become vital. For instance, in engineering manufacture components can be categorized as: critical, sensitive or ordinary. The classification is a function of the dependence of the project on the problems and importance of that component. An early concentration of the best staff resources on the critical components is one form of risk reduction.

The literature on concurrent engineering has so far given relatively little attention to its human side. The emphasis has been to shorten the lead time of technical and design activities. Human aspects were constrained by previous technical decisions and were handled in a fragmented manner. In a sequential process it was frequently thought that such matters could be resolved at the installation/commissioning stage.

6.13.2 The need for standardization

One major contribution to effective concurrent engineering is the standardization of as many components and sub-assemblies as possible. Equally important is the standardization of methods and procedure such as:

- product development procedures,
- component design methodology,
- manufacturing methods,
- quality assurance and test procedures,
- software packages, such as for simulation.

The use of company and external standards simplifies most operations, avoids much duplication of effort and reduces the risks of detail errors. There are many aspects of standardization, which will be dealt within Chapter 14. However, what matters with concurrent engineering is the maximization of commonality which allows development teams under time pressure to concentrate on the challenges of new technology and product concepts.

Company standards and manuals require a considerable amount of investment to set up and continuous attention to keep up to date. The standards manual can be in printed or in electronic form, where it can be readily available to technical staff on CD-ROM or via a local area network (LAN).

6.13.3 The role of key suppliers in concurrent engineering

When they select key suppliers many major manufacturers now insist that these have the capacity and capability to participate in the concurrent engineering of products and processes.

Both the suppliers and assemblers of key components, such as the manufacturers and users of electronic chips, have come much closer so as to compete with the lead time challenges of integrated manufacturers. Concurrent engineering involving such suppliers is becoming crucial, because of the rapid changes and increased sophistication in chip technology and the life cycle compression of the products in which these chips are used.

A particular development is the production of semi-custom chips, known as application specific integrated circuits (ASICS). It shows how flexible component design can be focused on product application opportunities. The chips have a unique set of circuits designed to customer requirements, but most of the chip is to standard design. This approach integrates some of the benefits of large volume production with a bespoke product.

In commercial terms, partnerships are established with key customers which make the component supplier almost an in-house source of chip technology. The supplier's chip designers work closely with the client's systems designers.

6.13.4 Rapid prototyping

The design specification for a major new product, such as an aircraft, can be a most complex and difficult challenge. To meet this a number of evolutionary design studies may be required before a suitable design solution can be found. On many aspects, design and development are interwoven and prototype components and systems have to be rigorously tested before their design confirmation can be accepted. Such work is both time consuming and very costly, particularly where a prototype failure has design repercussions on the specification of other systems.

The concept of rapid prototyping has been developed in the aircraft industry to cope with the design and management of complex systems. While an overall project has total design requirements, the effect these will have on the hierarchy of subsystems remains to be established, both for an individual system and the interaction effects of different subsystems. The design parameters at these levels have still to be confirmed and prototypes have to be assessed. Rapid prototyping is a method for simulating initial design ideas using commercially available components and computer software. Low-cost test rigs can be built, dismantled and rebuilt rapidly. The time saving avoids some of the risks when a manufacturer, under schedule pressure, has to start the production of the component before all design aspects are confirmed.

When the initial design idea is proved a prototyping tool is developed which is used for the building of the design that is then tested and evaluated. Only when the design concept is confirmed in this manner will the detailed design for production be carried out.

Rapid prototyping also allows computer software to be written, debugged, integrated and proven before the product is commissioned. This is especially important with aircraft where most of the controls and safety systems are computer-based.

A number of techniques have become established. For instance, in the car industry three-dimensional CAD systems are now integrated with vacuum casting equipment for the speedy production of prototype components. A stereolithographic model is developed from the CAD data and a silicone mould is cast around it. The mould is placed into a vacuum to remove air and then cured. The removal of the model leaves a cavity which is filled with a polyurethane resin. The resultant shape is the prototype component,

The uses of such prototypes are:

- Visualization: the designer can see and hold the component at an early stage of the design process. This helps to reduce obvious design errors.
- Assembly verification: the location of the component within an overall assembly can be checked quickly. This is important with limited space. The assembly process and the maintenance replacement task can readily be verified.
- Design analysis: the ready availability of several prototype components facilitates concurrent testing and simulation work thus reducing the lead time for component development.
- Rapid tooling: with the growing refinement and improved accuracy of CAD generated prototypes there is growing scope for the accelerated manufacture of production tooling which is often the crucial factor in the introduction of a new product.

6.13.5 The automation of development work

Where there is a need for large-scale testing and laboratory analysis a development section becomes similar to a production unit. Routine tasks, such as for drug testing, can be handled, day and night by computer-controlled test equipment linked with pick-and-place robots. Test results can be integrated automatically within established information systems.

The saving in development time and staff costs warrant this type of automation.

6.14 PARTNERS FOR NEW TECHNOLOGY DEVELOPMENT

6.14.1 The use of trade and research associations

There is a range of circumstances where a company may wish to contract out some of its R&D work. The following are the more important reasons:

- It wishes to commission special research projects for which it does not have the required human or physical resources. This is especially the case where it has no experience or skills in the field of research and it is not economic to enter it.
- Its research facilities are already fully committed and it does not wish to extend them.
- A small company may prefer to contain the scale, costs and risks of a large R&D staff commitment.
- The association may have special expertise in related matters, such as standardization activities.

A number of related aspects have also to be assessed. The company will have no detailed control of the project and it will be important to establish clear working arrangements to ensure that the research continues along agreed paths. Confidentiality and intellectual property rights have to be safeguarded. The ready transferability of the acquired knowledge can be critical and this can be assisted in some cases by the temporary secondment of key company staff.

6.14.2 The use of contractors for design and development

Depending on its main business activities, a company may contract-out the bulk of its engineering development work. For instance, a leading chocolate confectionary company manages the design and development of its own purpose-built high speed packing machinery. It has chosen this approach rather than buy standard packaging machinery. Apart from getting customized equipment to its own specification it achieves process differentiation and cost advantages in relation to its competitors. The bulk of actual design is contracted out to well-established design specialists with whom the company has had long standing relationships. Similarly, prototype development and fabrication is carried out by local engineering companies which are part of a well-developed supplier infrastructure.

The bulk of the company's high quality engineering staff is deployed on specification development and project management. Most of those engineers have had both engineering development and production line experience.

6.14.3 Collaborative research

With some fast-moving, advanced technologies, reliance on internally financed and conducted R&D is insufficient. The costs are too high; key staff are scarce and competitive pressures add urgency to achievement. This is, of course, one reason for technological alliances. Government pressure and support such as with the UK Alvey and the European Community Esprit programmes, have resulted in the establishment of research consortia involving a number of companies and in the latter case a number of different European countries.

Participation in a research consortium has repercussions on the company. There is the stimulus of discussion joint planning of large ventures, as well as the opportunity of some informal transfer of know-how. Dodgson (1989) also refers to the effects a consortium, such as the Alvey programme, has on the internal R&D organization of some of the participant companies. With some firms there is a radicalizing effect on its strategic thinking and more attention is given to longer term 'blue sky' research which is expected to grow into key R&D areas within five to ten years.

On the other hand, research consortia have also added their challenges to technical managers. Although the programmes are 'precompetitive', they have political dimensions and commercial implications. This requires diplomatic skills and the ability to work to common advantage with colleagues having a different language and culture. The development and confirmation of collaboration agreements have been a major source of delay. Basic principles could be embodied in a standard form of contract and specific, supplementary items left for agreement amongst participating parties would have reduced the problems of prolonged negotiations.

6.14.4 Links with other institutions

Even where some companies obtained most of their know-how from internally generated R&D they often have strong links with universities, research institutes and other industrial companies. Where they are embarking on a new technological trajectory, they often base their expertise on initial, externally acquired know-how.

Much can be gained, apart from possible grants, from close links with the relevant funding and government infrastructures. They can provide intelligence about longer term opportunities. The large corporation knows its way around the corridors of funding agencies; many small firms have no expertise or contacts to utilize such links.

6.15 REMAINING A GOING CONCERN

This and previous chapters have stressed the importance of competitive technology to a firm's prosperity and, indeed, survival. However, the quest for competitive technology must not be at the cost of every other activity. The company needs to remain a going concern, both financially and operationally.

When a major new product with new technologies is introduced it will often be alongside other products, the technologies of which may be more mature but are still undergoing incremental improvement. Their revenue producing capacity must be sustained to maintain cash flow. Where a new operational system is required for the new product it must therefore avoid causing undue disturbance to the existing production system until an economic and smooth changeover can be achieved.

6.16 SUMMARY

The realization of a new technology is based on research and development (R&D). Its commercial success is a function of development, design, manufacture and marketing (DDMM). Research policy, a subset of corporate technology strategy, has to be integrated with the overall business strategy. The management of technology intelligence is an important component of R&D.

The risks and problems associated with R&D need to be reflected in its management and organization structure. The syle of management, its flexibility and learning ability are important determinants of R&D success. Well-motivated, interdisciplinary project teams contribute to this.

The close integration of design with manufacture and marketing is a key feature of DDMM where competitive advantage depends on low-cost quality products, available on time. Knowledge-based systems and task automation, be it in design or development, use advanced technology for the more effective and quicker realisation of new technologies.

FURTHER READING

Amendola, M. and Gaffard, J. (1988) *The Innovative Choice*, Basil Blackwell, Oxford.

Beattie, C.J. and Reader, R.D. (1971) *Quantitative Management in R&D*, Chapman & Hall, London.

Heap, J. (1989) *The Management of Innovation and Design*, Cassell Educational Ltd, London.

Parsaei, H.R. and Sullivan, W.G. (eds) (1993) *Concurrent Engineering*, Chapman & Hall, London.

Roussell, P.A., Saad, K.N. and Erickson, T.J, (1991) *Third Generation R&D*, Harvard Business School Press, Boston, MA.

Woodward, J. (1965) *Industrial Organization: Theory and Practice*, Oxford University Press, London.

CHAPTER 7

The adoption of new manufacturing technology

7.1 INTRODUCTION

Chapter 6 was mainly concerned with the development of a new technology by the firm which was going to use it. It emphasized the close association between the new technology and the products which embodied it. This chapter considers the process of introducing a new technology to the firm and adapting it, particularly within the manufacturing function. The industries which are most affected by this are the engineering and allied industries; those which primarily make and assemble discrete parts and artefacts.

Many new technologies, such as in the field of production automation, are quite different to the core technologies employed by the companies in these industries. They tend to be knowledge-based. Their impact is not just the automation of specific processes; but the integration of automation systems, both physical and organizational.

It will be appreciated that many companies have to adjust continuously to technical change. For instance, replacement equipment or an updated control system often introduces new technology at the detail level and, although not always without complications, the repercussions are normally contained. Chapter 13 deals with the nature of such incremental innovation. This chapter is concerned with the introduction of new technology at the strategic level, involving major capital expenditures and reorganization.

A company can introduce an established, but unused technology that is well along its path of diffusion. The design, methods and know-how are confirmed. Experience and expertise are available to assist its introduction; particularly where an international corporation transfers the new technology from a factory abroad. The introduction is often easier and quicker but other challenges of the prospective change remain.

Alternatively, the technology may be completely new and in its early stage; much of the related methods, designs and details are still fluid and the company could be faced with a significant application development commitment.

The development of new technology in Chapter 6 mainly concerned staff whose job it was to achieve this. In this chapter the human side is different. It relates to factory operators, production supervisors, middle managers and other indirect staff who do not necessarily see technological change as their mission in life; who fear it, resist it and need to be persuaded to accept it and even participate in its implementation. This adds another dimension to the management of technology.

7.2 MANUFACTURING STRATEGY

The starting point for the introduction of a new manufacturing technology is the overall business plan and manufacturing strategy which form the basis of an integrated programme. The strategy is concerned with the adoption of new, as well as the disposal of, old technologies. With multiplant operations the latter does not always follow from the former. Typically, an international company may adopt a new technology in one factory but retain the displaced technology in other factories abroad.

The case for manufacturing strategy was well put by the *Financial Times* of 14 May 1985:

> Industry cannot rid itself of the 'work study syndrome' and its preoccupation with localised work improvement rather than overall systems: too much technology has been placed piecemeal in Victorian environments in preference to an overall plan for the manufacturing business.

The implications of this view are that not only has the purpose of new technology to be established but that the starting point is at the corporate level.

It is interesting to note, as a matter of policy, that new technology is not normally introduced in Japanese factories with the intention of reducing the number of employees. The aim is to increase the quality of the product or service, increase the product variety or to reduce production costs so as to reduce the product price.

7.2.1 The preparation for new technology

Where a major programme for a new manufacturing technology is envisaged, the time-scale can be several years. Such a programme normally contains feasibility studies, design evaluations and much detailed development and engineering work. The timing of staff consultations is also a relevant constituent of the planning activity.

According to Roth *et al.* (1992), reporting on their research investigations into 'world class' manufacture, competitiveness at the global level was less governed by the precise nature of the manufacturing technology in place, as by the company 'infrastructure' which supported that technology. The technology market is relatively open but effective transfer depends heavily on skill levels, human resources management, organizational structure, the emphasis given to quality and continuous improvement – in essence: company capability. If that is inadequate no amount of high technology will make that company competitive.

7.3 THE INTRODUCTION OF NEW TECHNOLOGY

'What is required?' The answer to this ostensibly simple question calls for a full and in-depth analysis of all the economic and technical aspects involved, so as to establish an accurate specification of operational and performance requirements. These will be within, and give expression to, the strategic framework. The analysis of existing operations must include and go behind the basis of those activities. Part of this evaluation includes the questioning of activities which have become enshrined with the passage of time; there can be candidates for elimination as well as modernization! The introduction of new technology when embodied in new equipment is not an instantaneous transaction. Whether it is computer-aided design (CAD) for the drawing office or a new type of automation in production departments, the introduction is mostly over a period of time. As Buchanan and Boddy (1983) indicate: technical change is a process, not an event. The introduction of new technology equipment involves a period of planning and preparation before arrival; adjusting, rectifying and improving it, once installed, and training staff to use and to maintain it. Even the effective introduction of relatively minor changes can be protracted if a number of parties are involved.

Clark *et al.* (1988) distinguish five stages in the process of technological change:

1. Initiation. The process by which managers identify and pursue an opportunity for the adoption of a new technology.
2. System selection. The process of in-house design and the development of a particular system or equipment.
3. Decision to adopt. This refers to the process leading to the decision to invest resources in the acquisition of the new technology.
4. Implementation. This embraces the process of introducing the new technology into the work place. This includes both technical and human aspects of installation, commissioning and debugging the new technology.
5. Routine operation. A stable, productive method of working has been attained.

7.3.1 Some requisites for successful introduction

Considerable attention has been given in recent years to the organizational requirements for major schemes such as the NEDO (National Economic Development Office) guidelines (1985) for the implementation of advanced manufacturing technology (AMT). The following are the more important guidelines:

1. There should be an executive director with responsibility for manufacturing strategy and operations.
2. Close liaison is important between marketing, design, production and finance, particularly on new product development.
3. There must be board level responsibility for research and development into products and production processes.
4. A nucleus of technically qualified staff, with time and resources, should undertake the preliminary planning and evaluation.
5. There should be early involvement of accountants when evaluating alternatives for new product designs and production processes.
6. It is very desirable to have communication procedures which allow for employees at all levels and trade unions to be consulted and informed on the company's competitive position and investment plans.
7. There needs to be a manpower planning system which provides up-to-date information on the company's available skills and experiences, as well as requirements for retraining and updating courses.

7.3.2 Feasibility studies

These initial investigations build on the first set of performance requirements. They also include preliminary financial assessments. Essentially, feasibility analysis is the first evaluation of an incipient proposition. It is concerned with the nature of technological risks, the anticipated development or application work, the order of magnitude of the required investment, the basic manufacturing costs and likely benefits arising from the adoption of the new technology. Manufacturing strategy furnishes some of the criteria by which the incipient proposition is assessed. This is the best time to consider alternative technologies and processes, as well as the level of the preferred technologies and systems. The feasibility study is often a relatively informal activity; significant organizational resources will only be devoted once feasibility is confirmed and the investigations given a 'project status'.

The feasibility analysis is normally undertaken by the functional department most involved in the area of operations, typically manufacturing engineering. However, even at this stage, it is already important to have an interfunctional team to provide stimuli and to test assumptions. A prospective proposition needs a chaperone, but also a second opinion;

somebody sufficiently apart from the proposal to have no prior view. The marshalling and tentative evaluation of all relevant information provides the setting for the feasibility decision. The main proposition elements and promises are related to the specified policies and criteria. The decision often combines technological and economic judgement. It may be made by the functional manager concerned or a technical committee respons-ible for such first-stage decisions.

The proposition requirements will be shaped into an investment pro-gramme with a cash flow pattern where, if possible, the yields from initial investments can help to finance the later stages of development. The first attention will be the 'quick hits' where a relatively good and quick return can be obtained, such as an immediate reduction in stock or work-in-progress. However, the temptation of a promising return must not become a hostage to the future, such as an island of automation which proves to be incompatible with subsequent developments. Furthermore, the needs of the ongoing business must always be met; managerial and technological staff time is never a free issue and must always be husbanded. A big technological jump forward can lead to uncontrollable second-order effects; measured evolution within a carefully planned and agreed framework can avoid many a crisis.

7.3.3 The adoption and development of a new technology

When a new technology is adopted the nature and embodiment of that technology correspond to its level of advance at the time of adoption. It is often the case that the technology is at a threshold stage and is rapidly developed further. The introduction of robots to manufacturing in the 1980s illustrates this. The first major applications were stand alone units used for simple but environmentally unattractive operations, such as paint spraying or die casting, or for component loading/unloading. Since then the degree of manipulation and interaction which robots have achieved has resulted in fully automated production lines, such as for the welding assembly of car bodies, where the robots are embedded within a highly integrated complex, computer-controlled system. The develop-ment of related tactile and optical systems now allows robots and their sensors to feel different material; to see shapes and manipulate them; to pick and place components for assembly; to recognize speech.

Sometimes such a rapid development poses a dilemma for the poten-tial user. At what stage should a new technology be adopted? Technological forecasting helps with the projection of technology trajec-tories. Adoption now is a commitment to a technology that may be rapidly overtaken. Postponement of the acquisition can leave the com-pany at a competitive disadvantage. Some companies have the policy of contained acquisition. Limited investments are made to contain risks and to acquire operating experience of the new technology.

7.4 THE CHALLENGES OF FACTORY AUTOMATION

Factory automation can be anything from a minor device of a conventional nature to a completely automated factory linked by networks to other factories and corporate headquarters. Automation is not new; 50 years ago it was mainly mechanical or electro-mechanical. Now microelectronics are the basis. Their flexibility, universality and relatively low costs permit the building of automated systems of a scale and complexity that revolutionize manufacturing systems. The challenge is to management and organization to harness the potential of microelectronic automation.

In the 1980s General Motors spent $80 billion in modernization worldwide. The corporation acknowledged that a proportion of that outlay spent on production plant, control equipment and computer systems was wasted. Waste occurred because of the inappropriate evaluation of needs and of financial returns, as well as an insufficient analysis of 'people systems' to which the Japanese devoted much attention.

Particular problems which were encountered consisted of:

- too much untried technology,
- advanced technology equipment which was too sophisticated for the tasks required,
- information systems which produced far more data than was needed,
- linked automation systems were too much hostage to reliability problems.

Too much automation and sophistication gave too many complications. With hindsight, an incremental systems approach would have been better than one big, ambitious automation programme. The situation was made more difficult by a lack of corporate standards and strategic investments.

A number of lessons were learned. It was better to start with manufacturing needs. Don't automate waste – eliminate it. Analyse the skill patterns of staff and workers who have to run the new systems. There will be a need for substantial training programmes. make visual management a tool. With a multiplicity of technological advances 'go lean'; make things simple; have no more sophistication than needed. The rate of assimilation of new technology is an important strategic consideration. Be flexible – go for compatibility and bolt-on systems.

These were the much-publicized experiences and recommendations of one of the world's greatest manufacturing companies with staff and financial resources equalled by few. The experiences of more modest companies have not had quite the same press but they can also add to the list of problems. The following are the more important:

- The unfamiliarity of many project staff with the subject matter of automation. The consolidation of experience is often affected by another technological leap on the next project.

- The time-scale of many factory automation schemes. While a step-by-step approach avoids some pitfalls, there can be a loss of direction and perspective.
- There are semi-development aspects with some installations, particularly with software systems.
- There is limited supplier application experience of the actual manufacturing systems requirements.
- The second-order effects of automation on non-automated/manual systems.

Generally, major automation schemes are based on computer-controlled systems which, for proper operation , require explicit rules and challenge the assumptions about how certain work is carried out. Systems audits are the preliminaries for automation projects and neglect here can yield a rich harvest of problems.

7.5 THE STAGES OF FACTORY AUTOMATION

Four levels of automation can be envisaged. Each has its own opportunities and challenges:

7.5.1 The automation of specific functional areas

These can be physical, such as machining lines, or computer-based organizational systems, such as quality assurance or material requirement planning (MRP) systems. Success is judged in terms of functional performance and for organizational systems in terms of data accuracy and its use by specialist staff for planning and control purposes.

7.5.2 Partial integration

At this stage several functions are linked via common computer files and coordinated control. Real-time updating becomes possible and management information becomes available for decision and control. General shop floor data collection is established.

7.5.3 Full integration

A common database serves the manufacturing unit. Local networks are established and information is generally screen-based.

7.5.4 Full automation

The total information system is now integrated with computer-aided design and manufacture, equipment operation, automated factory transport systems and automated test systems.

The following are the most important constituents of the integration process.

7.6 MANUFACTURING CELLS

This development in the field of engineering batch production has created groups of self-contained, integrated production units devoted to the manufacture of specific products or components, such as gears, shafts or the machining of complex castings. All the required equipment, irrespective of type, and all test equipment is located in the cell which is managed by a group of specific employees. Most of the equipment and the scheduling of the production programme is computer-controlled.

To obtain commitment to the changes due to new technology and to improve motivation, team operation is stressed. There is a premium on cooperation to achieve objectives and more attention is given to organization, social and communicating skills. Within the group, work allocation is flexible between different members and multi-skilling replaces traditional craft distinctions. There is a large measure of operational self-control and on the spot problem solving. Functional specialists give support: they don't control. The operators are responsible for their own quality. Some companies put seating accommodation and break facilities into, or adjacent to some larger cells. This facilitates discussion about the cell and its operation, such as production schedules or quality problems. The cell is often an integrated module within a computer-based information system but use is also made of notice boards which show cell performance in simple visual terms. Visual management is very much in evidence. Some Japanese companies encourage a form of rotating group leadership amongst the operators and provide in-house training for this.

Such cells are often 'islands of automation' within a factory. If it is intended, within an overall strategy, that the cells serve as precursors to computer-integrated manufacture (CIM) then care must be taken to ensure that the interfaces between the cells and the rest of the factory are compatible. This is not only a matter of systems but also of operational management.

Cells give more effective cost and work flow control than a traditional engineering layout. There are fewer boundary problems where a number of related operations are integrated within one cell.

Some companies have reorganized their factory entirely into self-sufficient manufacturing cells without taking the final step of total integration. Lloyd and Mason (1993) give a telling account of the transformation of the GEC ALSTHOM large electrical machines factory in Rugby. This major project physically and organizationally completely transformed operations and a traditional factory layout into 12 manufacturing cells. The project took five years, cost about £27M, required 130 man/years of planning and project management, involved the demolition and rebuilding of the factory – all this with continuous and increasing production on site.

The manufacturing cells were located to give a flow line layout, with the next cell designated as the customer. The company adopted the just-in-time (JIT) form of operation and delivery of materials from suppliers was direct to the cell concerned. The establishment of the cells was part of a 'renaissance programme' – everything was changed. Simultaneously, the company introduced total quality management (TQM), design for manufacture (DFM) a full CAD/CAM system and other state-of-the-art techniques fitting for a world class manufacturing company. Stage 1 of the project concentrated on a new factory configuration and manufacturing organization. Stage 2 plans to release the full potential of the work force, ensure a culture of ongoing improvement on the basis of a single status work force. By all measures of performance the project has been a success and ensured the survival of the company. An important contributory to this was an 'holistic' approach and coherent programme of which the establishment of manufacturing cells was an integral component.

7.7 FLEXIBLE MANUFACTURING SYSTEMS (FMS)

These are integrated machining centres, usually linked to automatic loading/unloading systems and automatic inspection. Transfer of the work in progress is either with pallets or robots. A centre has its own tool magazine and is programmed to automatically select its own tools and locate these precisely. Their key feature is numerical or computer control which allows the immediate setting of machines to suit different work pieces. The avoidance of major set-up times and costs makes FMS able to deal with very small batches and provides the flexibility of production which is their hallmark. To be economic, FMS requires effective tool and programme management.

Advanced manufacturing technology, such as FMS, is based on strong technological interdependence. For routine operations the degree of technological uncertainty has been reduced, as so much can be managed by the system itself, such as automatic compensation for tool wear. However, where an event cannot be handled by the system itself then there is a high degree of uncertainty. It also presumes an environmental stability; such as material available or tools in place.

7.8 COMPUTER-INTEGRATED MANUFACTURE (CIM)

As the name indicates, CIM is the organizational and physical control of manufacture by computers. Computer-controlled automation is applied to the physical processes and movement of work. It also provides continuous data and status reports as part of a departmental or factory information system. Individual machine reports are aggregated and formatted for management attention.

The establishment of CIM is a major undertaking. Its achievement requires a long term strategy, substantial expenditure and detailed planning over several years. Its ultimate form, the automatic and fully integrated factory, such as the one developed by Fanuc in Japan, has now been running for several years.

A number of companies have committed themselves to achieve full CIM and some are close to it.There are basically two ways to attain that objective. The first and the most common is the step-by-step approach which starts with manufacturing cells and then integrates these.The other is to introduce complete integration in one 'giant leap'. That approach may be on a green field site where there are fewer layout or industrial relations constraints. This is only feasible where a firm has already acquired the necessary expertise in existing factories. Either way is a challenge for the development of manufacturing systems which will lead to the complete transformation of production and its management. The human, training and organizational implications are profound and have steered most companies towards a step-by-step approach. For many companies the full utilization of CIM has still a long way to go.

Large and integrated databases are the key to a CIM system. To achieve effective CIM there needs to be a data strategy to define where and how data should be held and how it should interrelate so that all processors and procedures are compatible and integrated. Process planning determines what should be processed where, and what control system should operate. A communications strategy is the basis for physical and operational integration. This applies especially where manufacturers of automation equipment use different communication systems which do not always interface with each other. The choice of a suitable local area network standard, such as the General Motors Manufacturing Automation Protocol (MAP), is crucial.

Successful CIM will involve a number of functional areas such as design, process planning, production control and inventory management. Its role will affect the total operating system, from the sales quotation and order entry to the dispatch of the product. The control system will be managed by a network of PCs and a master database. Integration is vertical from top management to an operating unit on the shop floor. It will be flow structured from the first customer enquiry to the dispatch of goods and the payment for these. It will also be lateral to include parallel manufacturing activities. A particular effect will be the rapid exchange of data, such as reports and schedules between departments, which allows much faster decision making and responses to contingencies.

Manufacturing cells, FMS and CIM require very task-specific training. Many companies have found that such needs are best met by in-house courses or specially tailored training programmes.

7.9 COMPUTER-AIDED DESIGN AND MANUFACTURE (CAD/CAM)

Computer-aided design (CAD) is a well established facility in most design offices. The universal use of personal computers (PCs) and the wide range of software packages permits the ready development of design configurations, drawings and graphic displays. Linked with the appropriate databases it provides bills of materials, cost data and contract documents. Computer-aided design has increased significantly the productivity of many design offices. The design-development lead time has often been shortened, yet self-checking routines have made design analysis more effective.

The essence of CAD/CAM is the combination of design with the preparation of machining programmes for computer-controlled machine tools and inspection programmes for computer-controlled quality assurance measuring equipment. Through the integration of design and manufacturing planning data the detailed manufacturing tasks can also now be optimized in terms of such criteria as throughput times or tool life. An important current development is the feedback of manufacturing data, such as from quality assurance, to allow the designer to incorporate the realities of equipment performance into their design decisions. The links between the functions of design and manufacturing planning have become very close.

According to a major UK equipment manufacturer one significant advantage of CAD is to be able to generate design drawings and machine tool programmes from the same database. The company's adoption of CAD/CAM was 'by reasonable steps'. For its business and the type of components it handled two-dimensional CAD was appropriate. The sophistication of three-dimensional modelling was regarded as too costly and complex.

7.10 INTELLIGENT MANUFACTURING SYSTEMS (IMS)

This is seen by some as the ultimate in manufacturing systems development and enterprise integration. It is the next generation extension from such systems as Manufacturing Automation Protocol (MAP) and integrates the total product process, from R&D to manufacture and marketing to eventual customer satisfaction with the product. The objective is globalization of flexible manufacture at world class level. A good account of IMS has been published by the UK Institution of Electrical Engineers (1993).

IMS is a Japanese initiative with ambitious objectives. Its research programme is envisaged as a collaborative, international initiative with Japan, the USA and the European Union as the main partners. The following are the key areas for development:

1. Production system development concepts: such as system architecture, integration techniques, systematic production/quality control technology, evaluation methods etc.
2. Production-related information and communications technologies: including the development of global databases, operating systems and network utilization techniques.
3. Production/control equipment and processing technology; concentrating on high-level intelligence machine tools, robots, control and processing technologies.
4. Application technology for new materials: such as composite materials and ceramics, to achieve major performance improvements for production plant.
5. Human factors in production: such as the development of new production environments and human interface technology.

Considerable preparatory work has already been carried out in Japan as part of a two-stage domestic feasibility studies programme, comprising 45 projects. The Japanese IMS Promotion Centre, support by the Ministry of International Trade and Industry (MITI), is the focus and coordinating body for this scheme. Its company membership in 1992 consisted of 68 core and 20 supporting members as well as 28 universities and public research institutions.

Many of the component technologies already exist: the challenge now is to form them into a coherent system. A crucial requirement will be standard communications interfaces. IMS is an all-embracing concept where the costs, scale and risks of a constellation of major research projects warrant international alliances. Whether these will come about depends also on political and economic factors, not just technological aspects.

7.11 THE OPERATION OF THE NEW TECHNOLOGY

When a new technology is introduced it will be part of the implementation process to assess how it will be used once it is operational. Decisions will be needed about the task content of work; how it will be supervised and controlled. The pattern of work organization and the process by which it becomes established is a key determinant of work performance.

The installation of automation equipment has design implications. For instance, a welding robot can weld more consistently than many human operators. With welds of a higher quality and consistency it will be feasible to cut or rearrange the number of welds that are required, say, for car body assembly. Process planning for a robot is more than the transposition of the agent of production.

Where robots and new computer-controlled plant are installed there is often a need for a higher standard of housekeeping and cleanliness. This has other opportunities for improving the work environment for operators. Hard, dangerous and dirty work is often eliminated.

Some analysts suggest that existing organizational structures influence the adoption and operation of a new technology. Problems with the installation of FMS and assembly automation are often due to ultimately organizational rather than technical problems.

7.11.1 Preparation for integration

Computer-based integration requires certain attitudes; 'pegboard thinking' as someone has described it. It requires a sense and quest for order, logic and consistency. If that approach is not developed by all associated staff, not just computer specialists or systems engineers, but also by operators, clerical staff, technicians and managers, the integration project will have difficulties. A manager may accept your ambiguity: your computer won't. The audit, simplification, systemization and improvement of working procedures is a precondition of integration success. With some companies such preparatory work has yielded much of the expected improvement without even installing the computerized system.

Procedures need to be proven at module level but then must also match on integration. The mere addition of modules does not capture the synergy of integration. There is a need here for experienced systems integrators who can effectively link plant to information systems.

Many computerized systems fail because the companies implementing them are not looking at the overall structure of, and information flow within, their firms. Computer-aided design and manufacture (CAD/CAM) systems, for instance, are often introduced as a set of unconnected, closed systems without enough attention given to information transfer. The two systems are essentially islands of automation. Their integration and the possible future addition of corresponding quality assurance or costing systems will be correspondingly more difficult and costly. The repercussions on the organization structure need to be resolved with greater flexibility to allow growth and change to take place.

7.11.2 Initial operational problems

High technology equipment with limited application exposure is prone to reliability problems. There is the debugging of software systems and the settling down of interfaces. Some control units may not be rugged enough for the operational environment. This puts a premium on improvization skills and quick learning.

Again, with the use of robots, problems may occur not with the robot, but with the ancillaries transferred from traditional equipment, such as fixtures, clamps etc. They can impede the easy manipulation of robot arms.Also, the invariant nature of some robot operation cannot completely mimic the flexibility of human operators where dimensional tolerances are generous or there are problems with surface finish, such as with burrs from previous machining operations. The introduction of a robot into a production flow line often has both upstream and downstream repercussions. The line may have to be rebalanced, i.e. the work pattern has to be rearranged.

7.11.3 Some problems of integrated systems

The design and installation of integrated systems is already a major challenge. Their maintenance and effective operation makes further demands on staff and management. Integrated systems often have both automatic data and human inputs. The limited reliability of the latter is often the source of mischief. The following are some typical operational problems:

- the fragility of input data: do the reported stock figures actually match the stock in the stores?
- forgetfulness: staff may overlook putting key data into the system;
- misbookings: accidental or otherwise;
- incomplete inputs: for instance, process routes have changed but some operations have been missed or a tool identification code has not been updated;
- the codification of improvization: an emergency routine is not deleted when the emergency is over. High cost process routes become permanent;
- time lag in getting information on file: substantial delays were experienced by a precision engineering company with goods inwards procedures;
- staff can develop a 'decision fear' when swamped with information;
- misleading simulations; when planning assumptions don't match shop floor realities;
- the system is hostage to the hardware configeration; what happens when the existing mainframe computer has to be replaced;
- undue reliance on key staff.

As the degree of integration grows within a manufacturing system it may reach the stage of diminishing returns. Additional benefits are harder to attain while the additional integration costs tend to grow. Total integration is not an economic concept as such; its net benefits must speak for it. It is part of investment evaluation to determine the most effective level of integration for the business as a whole.

7.11.4 The development of open systems

The development of computer-based networks and the application of protocols for standard communications, such as the Open System Interconnection (OSI) model, extends a company system to suppliers and customers, be they local or anywhere in the world. Information can be delivered and processed through high speed communication networks. Within a company, different networked systems for design, finance, manufacturing or stock control can operate jointly and simultaneously, irrespective of the precise location of the function. For instance, a move in currency exchange rates can actuate the shift of a production order from one country to another within seconds! An order comes through the electronic mail and so do all the successive documents of the transaction. A company which can only accept orders through the post is at a disadvantage.

The important aspect of a company's technology strategy is to be able to secure the option of coming into such systems without undue costs, if and when its business situation warrants it. Compatibility of systems and equipment, through the use of standard systems, is a necessary requisite.

7.11.5 The integration of systems from different suppliers

A medium-sized engineering company had particular problems with the low utilization of its machining centres. Four of the centres and their control systems had been supplied with different guarantees by the manufacturers concerned. Breakdowns of the integrated systems became extended because of the disagreements with and between the various suppliers as to respective liabilities.

There is also the problem of a company having different software running on different computers. Initially, one firm chose its systems for functional objectives, such as production control with MRP. Over time it acquired a collection of incompatible systems. The installation of such systems and the achievement of their expected performance is usually costly in staff time. Their databases constitute a substantial investment and the 'sunk cost' which they represent is a discouragement to further systems development if they are in unsuitable formats.

7.12 CHANGE MANAGEMENT

Change management can be seen at three levels. The first is a total, step-function change of the company's operations. This is strategic, operational and cultural. It affects all members of the firm. The second is at a functional/department level, although even here the ramifications of a change can affect operations in other areas. The third is concerned with the management of improvement where change is incremental and adaptive.

Here, change management is considered at the second level, associated with the adoption of new manufacturing technology. Its introduction, embodied in new plant and processes, is both a step function and an adaptive change. The continuous pressure of technological and other changes has prompted some companies to introduce organizational responsibilities for the management of change. The management of change needs a professional approach and different skills from that of ongoing operations. The specific setting of any particular change is usually in the province of particular groups of functional or departmental managers but there are risks if change is function dominated. The analytical skills, breadth of view and experience related to the management of change are often best harnessed within task groups. In many ways change management is a form of project management with its own set of principles.

The first requisite is a set of clearly established objectives and priorities which have the continuous support of top management. Secondly, there must be a clear individual responsibility for the formulation and implementation of the change programme. This responsibility includes the process of consultation with all affected parties so that these have a clear understanding of the nature and implications of the changes to come. They must share and understand the vision. There is a case here for the 'socio-technical' approach – a balanced approach which considers technology, structure, tasks and people. Much of the benefits of technical change can be lost without a change in attitudes and organization.

A socio-technical system comprises two independent but related parts; the technological part which consists of machines, tools and techniques and the social part which consists of people and the relationships between them. The socio-technical systems are open systems which exist in a larger environment. Work design is aimed at jointly satisfying technical requirements and human needs within that environment. It must deal with the 'what about me?' questions. Goal congruence is important for the successful implementation of major changes.

There is a technical interdependence which expresses the degree of human cooperation required to satisfy technical requirements. There is also a technical uncertainty which governs the amount of information processing and decision making an employee must do to carry out his task. The amount of technical uncertainty determines whether operational control should be carried out externally, such as by supervision and or detailed procedures, or by the employee. With a high degree of uncertainty the work is best designed for employee self-control.

7.12.1 Behavioural aspects

Buchanan and Boddy (1983) point to the different perceptions and expectations of technological change by different levels of management. Technological change is seen as a political process as the expectations of those concerned often conflict. Changes to the organization that accompany technological change reflect strongly and directly the expectations and objectives of management and weakly the characteristics of the technology. The consequences of such change depend on the goals, assumptions and values of those who make the decisions. The capabilities of technology are enabling rather than determining.

Kanter (1983) has provided a useful insight into the behavioural aspects of change management. In the first instance organizational tools are needed, typically for:

- information – data, technical knowledge, expertise and political intelligence,
- resources – funds, materials, space and time,
- support – endorsement, backing, approval.

Change projects are easier to sell if they can be demonstrated on a pilot basis, also if they can be made tangible and divisible into easy stages. A project which is 'reversible', so that a return to the *status quo* is possible, if necessary, may be more acceptable. So is a change which fits in with the overall intent.

As part of the preparation for a change programme Twiss and Goodridge (1989) have developed the technique of 'force-field analysis' which is a semi-graphical audit of what and who will help or hinder such a programme. 'Stakeholder analysis' is a similar technique with an organizational behaviour approach which focuses on individual/departmental gains or losses of power or benefit.

7.12.2 Some practical aspects of managing technological change

Where one technology replaces another it is common for some techniques and know-how to become redundant. There is often an associated loss of standing, the challenge of adjustment and a significant learning task to come. Expertise in a displaced technology is not necessarily a qualification for the next technology. This engenders uncertainty, fear and resistance. The experience has generated much advice to managers on how to handle technological change, particularly as to the manner of bringing it about. Perhaps the most succinct is given by Buchanan and Boddy (1983) whose recommendations include the following:

1. Avoid technology-driven change. Concentrate on problems or opportunities, not technological solutions looking for problems.
2. Don't assume that human problems are bound to be solved by technical solutions
3. Check that the change can be justified in terms of improvement in overall performance.
4. Assess the effect of technological change, particularly automation, on related work areas which ostensibly do not seem to affected.
5. Consider the changes in information flow, resulting from the technological change and anticipate related changes in political power.
6. Identify the effects on job security, payment systems and career opportunities. These aspects are more likely to generate resistance rather than the changed technology itself.
7. Calculate carefully the time it will take to install and commission the new technology and make heavy provision for likely overruns.

Where a technology change is due to a technology transfer from a foreign company a participatory approach is particularly important. Seconded staff from the transferring company are often key suppliers of process know-how. Their acceptance by the work force and their integration within company teams is an important condition for successful technology transfer.

7.12.3 Implications for organization

There is a problem with the 'management by objectives' philosophy if a manager is judged by his ability to improve his specific area of performance. He focuses on his function and his department and this leads to a tendency for the proponents of new technology to think primarily in departmental rather than business terms. With advanced manufacturing technology the benefits generally accrue across departmental boundaries. Manufacture has to be redefined in overall work flow terms and a departmental manager will have to be judged on his capability to serve his customers, be they external or within the company. The role of the departmental manager can be critical; so much depends on how innovative he is.

The integration of both technological and organizational design should match strategic requirements. Also, it is important to remove the barriers of perspective and discourse between those responsible for strategy, technological development and work organization.

The design and implementation of technological and organizational policies is a social process, the success of which is affected by the perceptions and interpretation of those involved and the political skills and general competence of those responsible for them.

7.12.4 The role of supervision

As we have already seen with manufacturing cells, automation and advanced manufacturing technology have changed the pattern of work groups and greatly affected the role of supervision. 'Human' tasks such as the provision and analysis of process information have become automated. This has changed the role of the supervisor who is now required to:

- exercise leadership and provide guidance in terms of the new technology;
- develop skills of anticipating and solving a wide range of problems;
- be more concerned with the development of teamwork within a framework of self-determination and autonomy of work groups.

He is more of an adviser and less of a disciplinarian. As the scale, sophistication and capital cost of the equipment increases the supervisor is more a unit manager. A higher level of general and technical education will be required. Leadership qualities are essential. Personal and interpersonal skills will be needed to handle a wider spectrum of problems. The new technology line or manufacturing cell often represents a key investment by the company and correspondingly calls for a key manager to be responsible for it.

7.13 PEOPLE AND TECHNOLOGY AT WORK

When a new technology is introduced a technical specification is required. Corlett and Richardson (1981) hold that in the same way a social specification is needed for the resultant plant or office. Trends towards better performance, safer and more satisfactory jobs are linked to more interactive planning and management. The detailed knowledge of each stage of production cannot be entirely with management; experience of working the process is vital knowledge. A greater involvement in the design of the job by the person who will carry it out arises naturally from a concern with efficient and humanely designed work.

The design of work systems is often based on simplistic or mistaken assumptions about human behaviour. The needs of people from work tend to be neglected and human attributes are insufficiently considered. In the management of work systems there often is insufficient feedback about operational problems and implementation difficulties. For instance, employees with greater social and development needs may be more responsive to work in self-regulating groups and enjoy the challenges and opportunities of work with advanced technology. There is benefit in the encouragement of voluntary self selection.

7.13.1 The reduction of change-induced stress

Stress is an emotional, physical and cognitive reaction to a situation which places special demands on an individual. Most people function best under conditions of moderate demand. if the level of demand is either excessive or insufficient an imbalance is generated, which leads to stress. Cox (1978) makes the point that stress arises when there is an imbalance between perceived demand and the person's perception of his capability to meet that demand. Stress can have a range of causes, some of which reflect the characteristics of the person, his or her private life, as well as the social, physical and task demands of work. Uncertainty and change over which the individual has no control are sources of stress which expresses itself in anxiety, frustration and aggression; i.e. in emotional terms. Lazarus (1976) sees stress as an interactive process. It occurs when there are demands on the person which tax or exceed his adjustive resources.

The changes due to the introduction of a new technology can therefore cause stress and a perceptive management can do much to reduce it along with the resistance-focused behaviour which may result from it. It can allow for adaptation and coping with the new situation.

White (1984a) provides a helpful set of suggestions:

1. Do not change everything at once. Leave a stable and secure basis from which new arrangements can be explored. Leave work teams together if they are working well.
2. Ensure that there is adequate and direct feedback about new methods.
3. A step-wise, incremental sequence of changes, giving adequate time for acclimatization, is worthwhile, with identifiable achievements signalled and publicized.
4. Conflict and resistance are likely to be reduced if an open, exploratory style is adopted, encouraging collaboration and help seeking, looking at the causes of mistakes rather than looking for someone to blame, looking at how work is actually done rather than the procedures originally laid down.

White also makes the point that the very process of change provides the opportunity of review of the person/job fit and the reduction of conflict.

7.13.2 Participation in technological change

A report of a study tour of some Japanese factories by the work research unit of the UK Department of Employment noted that those companies involved their work force more and were more inclined to listen to and act upon their suggestions. New technology was seen by management and workers as one way of improving the quality of working life as well as enhancing the effectiveness of the enterprise.

The European Foundation for the Improvement of Living and Working Conditions (1992) in its survey on the attitudes of managers and employee representatives to employee involvement in technological change, interpreted involvement as:

> any participatory procedure or practice, ranging from the disclosure of information to consultation, negotiation and joint decision making, which formally or informally involves the parties concerned with the introduction of new technology in the discussion of decisions concerning the process of change.

The Foundation viewed the introduction of new technology as passing through the following phases: planning, selection, implementation and post-evaluation. It focused on the strategic planning and the operational implementation phases and found that the involvement of employee representatives at the planning stage was very low. At the operational level there was a considerable increase of consultation and joint negotiation. Consultation was more concerned with matters of direct interest to the work force, such as work structures and health and safety.

Managers generally were aware of some of the benefits of participation. It rarely slowed down the decision-making process. This was often because points came up that were relevant to a decision but had not been previously considered. Industrial relations improved; identification by the work force with company goals was enhanced; it facilitated the use of existing in-house skills and increased the acceptance of new technology. This was particularly important where management depended heavily on the skills and cooperation of its work force to introduce new technology.

An illustration of a successful participative approach to the introduction of new technology is given by Osola (1986) The new technology was the automation of cylinder head assembly at Perkins Diesel, a major manufacturer of diesel engines. The first important aspect was ongoing communication to ensure that every employee had a realistic understanding of the company's position in the market place, the competitive pressures within which it had to operate and the long-term company prospect.

Participation was linked with training opportunities. The company invited applications for jobs on the new line and training for those selected started months before the new equipment arrived. Part of the training included a week with Fairey Automation to see the new automation equipment as it was built. This elicited a number of suggestions by the prospective operators. Although minor in nature, but useful, they increased operator involvement and understanding of the equipment.

Participation in the introduction of new technology generates a feeling of ownership of that technology when it is successfully adopted and works. As White (1984b) indicates, the participative strategy for the management of change pays off because:

1. It affords opportunities to learn within the organization, to live with the new system and to cope with change as part of the way the company normally functions; joint learning by doing results in greater technical and social competence within the groups affected by change.
2. It involves people as participants in, rather than as passive recipients of change.
3. The process of 'selling' the new system, of getting acquainted with it and its effects, is not a separate issue but an integral aspect of change.
4. It helps to identify those matters about which negotiation between the management and employee representatives is needed.
5. It establishes resources within the organization which can be used to future advantage.

One benefit of effective participation is the feedback from job occupants to systems designers, particularly at the commissioning stage of a new system. This can refer to:

- Non-routine events, i.e. those which were not anticipated;
- queries generated,
- bottlenecks and interruptions to the work flow,
- human interactions in the work process,
- task interdependencies.

7.14 WORK STRUCTURE

The 'one person, one job' view of classical scientific management, which is the basis of so many organization structures, is being increasingly challenged by flexible, computerized systems. These systems integrate work flow and provide the opportunity to examine how jobs relate to each other in a changed setting.

7.14.1 Work tasks, job content and skill

According to McLoughlin and Clark (1988) the content of a job can be expressed in several dimensions. The most common is in work tasks. This describes the range and variety of the required duties, their complexity and the effort levels required to carry them out. Another is the autonomy the job holder has in executing his work; i.e. the degree of control under which he operates. A particular feature is the skill content of work which can be assessed in terms of the technical complexity involved.

New technology can be used either to eliminate the need for human intervention or it can be applied to support it by deepening human skills. It can provide a job challenge giving variety and the opportunity to learn. For instance, some engineering companies have enriched machine shop work by teaching operators to programme numerical controlled machines.

7.14.2 The nature of the work task

The introduction of new technology, whether automation or not, will affect the human task content of production. With most automation investments this has been one of the main points of economic justification. But there is a limit, a trade-off line, where it is not necessarily economic to replace every manual task. There may thus be a bundle of tasks which are automation determined and other activities of a traditional nature. Their combination, to ensure the effective deployment of a restructured labour force, needs to be established. White (1982) has summarized the main aspects which such a review should take into account. In his view, tasks should:

- combine to form a coherent job, either alone or with related jobs, the performance of which makes a significant and visible contribution;
- provide some variety of pace, method, location and skill,
- provide feedback on performance, both directly and through others;
- provide some degree of discretion in carrying out tasks, responsibility for outcomes and, in particular, control over the pace of work.

Where the operator role is essentially that of a monitor the job design should be such as to reduce the errors of monotony. Boredom induces stress.

The continuous feature development of automation equipment such as computer numerically controlled (CNC) machines, has increased automatic monitoring and fault diagnosis. 'Fire brigade' monitoring by operators has diminished, allowing a greater level of multi-machine supervision and the carrying out of tasks independently of machining cycles.

Overall, manufacturing automation has reduced the number of machine operators, material handlers and expediters, but has increased the number of planners, computer support staff and training specialists.

7.14.3 Job redesign

It has been observed that new technology provides a wider choice of different work practices. A number of changes become feasible. It allows staff to become familiar with the very process of change and, if this is well handled, will make future changes more acceptable. An example of such a choice is the programming for CNC machines in engineering factories:

whether it is the operator or a special programmer/technician who writes the programmes.

A further change has been the introduction of elementary maintenance tasks into production jobs. New developments in electronic testing facilitate self-diagnostic fault finding which greatly simplify fault tracing and reduce search time. Where this is coupled with modular components it will be feasible, in skill terms, to transfer such tasks to production operators. More complex problems can then be left to specially trained maintenance technicians. With major, intractable problems the equipment supplier or specialist contractors have to be called in although the costs of such support often make this the last resort.

With due regard to the technological aspects of a new operational system the work design task should be integrated with it. The implementation of a new technological system tends to be mechanistic but work design is a more developmental process and requires a longer lead time because of consultation, the skill learning process and feedback from work groups. The human aspects of work design have been stressed by Cummings and Blumberg (1987) who emphasize two kinds of personal needs. First, social needs; i.e. the desire for significant social relationships and second, growth needs; the desire for personal achievement and development.

Insufficient attention to these aspects can lead to human problems, expressed in terms of: pay, performance, such as each shift doing its jobs properly and not leaving things to the next, safety and equipment maintenance.

7.14.4 Effect on employee skills

With technological change, management has the choice of diminishing or enhancing the skills of staff. In some cases of automation there is a complete loss of skill, which in the extreme case leads to the redundancy of operators.

There is a crystallization of employees into:

- specialist key staff with the skills to use and to develop the new technology. Especially if this is a key technology, such staff will enjoy enhanced standing and conditions of employment.
- staff with an obsolete skills spectrum, the value of whose skills has declined. These have less stable jobs. They could be retrained to specialist key staff level but that is a matter of demand and retraining costs
- unskilled staff

Do you enhance or degrade the skills of your shop floor employees? With advanced manufacturing technology (AMT) a downgrading strategy would transfer many detailed operational decisions to supervisors and unit managers. This gives shop floor management more control but could increase indirect staff costs. There is also the need to pick up operating problems not reported automatically by the system which an operator could detect.

Child (1984) summarizes the aspects which the choice of employment policy needs to consider:

1. the amount of non-programmable skill and judgement which is utilized in the performance of tasks;
2. the degree to which the work being done contains an element of personal service;
3. the extent to which the pattern of work required is unpredictable;
4. the risks involved should the new technology/system break down.

The more these aspects are present the more appropriate would it be to regard the new technology as enhancing rather than replacing the skills of employees.

7.14.5 The redevelopment of skills

Preoccupation with daily operational problems can lead to a lack of attention to the required spectrum of labour skills when new technology is introduced. The sudden realization that key skills are missing causes firms to scramble to recruit skilled staff or to resort to improvised short-term solutions such as raising salaries and overtime payments to staff able to work the new systems. A greater emphasis on work force issues at an early stage may well help to overcome the problems of recruitment, skills shortages and increased training costs which appear to be the by-product of introducing new manufacturing technologies.

As the rate of technical change increases training becomes a more important and time-critical task. More volatile business conditions and the flexibility required to survive in such an environment calls for a greater range and depth of skills. Staff need to be able to change roles more quickly and to use their initiative. Skill training is most effective when given within the context of the business so that trainees can better envisage the effects of their performance on that of the firm. Self development is a feature of the training in the definition of which staff should participate. This has been emphasized by a number of Japanese companies where the introduction of new technology has meant that operators would have jobs requiring higher levels of skills following the installation of new equipment. The emphasis is always on training, so that the operators could carry out the new jobs that were being created.

Training has been defined by the Training Services of the UK Manpower Services Commission as 'a planned process to modify attitudes, knowledge or skill behaviour through learning experience to achieve effective performance in an activity or range of activities'. It is interesting to note that training is more than just the imparting of knowledge or a skill, and that attitude is given first place here.

One of the likely benefits of employee participation in the very planning of training is the improved design of training courses. Discussions about skill development can lead to more effective learning and often shorter, more economic courses.

Training courses need to have objectives, structure and agreed methods. They are needed not only for operators but also have to be designed for the various levels of management. This is most important where the success of the new technology depends on a change of culture. An open form of leadership and a participative style of management is a gift to some but a necessary acquisition for most. Consultation and participation demand particular personal skills which many managers have to learn.

The content and amount of training required depends on company circumstances. There will be the general need for training programmes to provide employees with the required multiple skills to detect and control technical and environmental variances and the social skills needed for group problem solving. Senior managers need to be exposed to best practice operations wherever these are accessible. Specific training will be needed for the application of CAD/CAM, TQM and other techniques. Everyone gains from awareness seminars.

An important ingredient of any course is the preparation of personal action plans by every course member, specifying what each will do. This is then followed up by a review meeting at a later date to see whether the actions had been taken and what lessons could be learned from implementation.

7.14.6 Job evaluation

Job evaluation is the process of analysing and assessing jobs in a systematic way so as to establish their relative worth for payment purposes. It is essentially based on a framework of existing organization and operation. Where new technology changes work patterns and job content there will be a lack of fit. Jobs will have to be respecified and revalued by a new technology job audit. What will also be important is that the criteria of value may have changed. Where the work is more on a group basis, as with a manufacturing cell, it will be more difficult to define specific individual tasks.

With new technology, job descriptions will often show a major switch in emphasis from physical and manipulative tasks to the perceptive and interpretative skills of an essentially monitoring role. This will affect work factors, grade descriptions and comparative weighting. Job specifications have to be rewritten with much more emphasis on mental tasks, rather than physical effort.

Of course, job revaluation is not confined to operators but is also relevant for supervisors and some levels of management.

7.14.7 Payment systems

With many introductions of new technology, piecework or other well-established bonus systems no longer match requirements and have been abandoned. Pay systems need to be adjusted to revised work structures, new forms of organization or increased levels of responsibility. A number of companies have put operators on a staff grade with salaries which have consolidated or enhanced previous earnings.

The acceptance of change is conditional on a reward system which, in the end, is seen to be in the interests of all involved.

7.14.8 Aspects of negotiation

The many changes of work structure and the very transformation of whole factories has involved some companies in continuous negotiations with its work force. The nature of negotiation varies with industries. In some cases the management/workforce dialogue has been with elected worker representatives, in others with trade union officials. Some companies deal with unions individually; others insist that they will deal only with one union. In the 1970s and 1980s a number of unions and companies established new technology agreements.

The details of negotiations are outside the scope of this book. What matters here is that negotiations, irrespective of the precise partner, take into account the realities of the changed technology and its repercussions on the human community which is affected by it.

7.15 SUMMARY

The adoption of a new manufacturing technology is the fusion of corporate technology and manufacturing strategies. Manufacturing automation plays a major part in new technology projects and has transformed engineering production. Its advent, however, has caused problems of organization and adjustment which call for a different approach to management.

The integration of diverse production operations has generated several stages of factory automation, such as manufacturing cells, flexible manufacturing systems, computer-integrated manufacture, computer-aided design and manufacture and, at the apex, intelligent manufacturing systems.

The manifold problems encountered with the introduction of new manufacturing technology and system has generated the concept of change management. This considers the 'socio-technical' implications of the new technology, the resulting changes in work systems and skill requirements and the reduction of change-induced stress.

FURTHER READING

Buchanan, D.A. and Boddy, D. (1983) *Organizations in the Computer Age*, Gower, Aldershot.

Child, John, (1984) *Organization: A Guide to Problems and Practice*, Harper and Row, London.

Clark, J. *et al.* (1988) *The Process of Technological Change*, Cambridge University Press, Cambridge.

Corlett, E.N. and Richardson, J. (eds) (1981) *Stress, Work Design and Productivity*, John Wiley & Sons, Chichester.

McLoughlin, I. and Clark, J. (1988) *Technological Change at Work*, Open University Press, Milton Keynes.

Voss, C. (ed) (1992) *Manufacturing Strategy*, Chapman & Hall, London.

Wall, T.D. *et al.* (eds) (1987) *The Human Side of Advanced Manufacturing Technology*, John Wiley & Sons, Chichester.

Williams, D.J. (1988) *Manufacturing Systems: An Introduction to the Technologies*, Chapman & Hall, London.

Wu, Bin (1992) *Manufacturing Systems Design and Analysis*, Chapman & Hall, London.

CHAPTER 8

Project management for new technology

8.1 INTRODUCTION

Project management is essentially an executive task. In this context its objectives are:

1. to formulate a meaningful proposal for authorization by the board of the company;
2. to implement the authorized project.

This chapter relates the accepted principles of project management to the specific aspects of both R&D and the new product development, covered by DDM (design, development, manufacture and marketing). The development of new technology includes both the embodied form, either in new products and/or new processes, as well as the disembodied form, such as a new computer-based information system.

The chapter is closely linked with Chapter 9 'Investment in New Technology' as project management here includes the formulation and implementation of what is essentially an investment proposition.

In project management terms the main feature of a project is that it is limited in duration. A manager is appointed and staff is chosen; when the project has been completed its personnel revert to other roles. This essentially temporary organization has implications for project management and the affiliation of members of a project team.

There is an extensive literature on the principles and practice of project management and those who wish to take the general subject further are advised to consult the updated texts of such writers as Lock (1992) or Harrison (1992). However, relatively little has been written about the application of project management to R&D and DDMM. With the inter-functional nature of projects in these areas, there is at least as much of a need here for project management as, for instance, with a major construction project. The purpose of this chapter is not to repeat what has already been well documented, but to add the context of new technology to the subject.

Apart from achieving the set project objectives within the given time and budget targets, R&D projects have the challenge of accomplishing technical performance requirements. By their very nature such projects deal with technical unknowns and the associated problems often lead to time and cost overruns. Much of the work, for instance, of an engineering development section is akin to a small jobbing shop. A small quantity of a wide range of tools, fixtures and materials is often required, the lack of any one of which can cause a project hold-up. There is therefore a premium on forward planning and organization which require the skills of project management. Continuous improvization is not enough.

8.2 THE BASIC CONSTITUENTS OF PROJECT MANAGEMENT

The extensive subject material can be assembled within the following major constituents:

- project organization
- project preparation
- project systems
- project cost management
- project time-scales
- project risks and changes
- project control
- project audit.

Some aspects, such as appropriate management and leadership styles have already been discussed in Chapters 6 and 7. The remainder of this chapter will focus on some of the specific project aspects of new technology development.

8.3 PROJECT PREPARATION

A number of tasks need to be carried out at this stage. Most of these are part of normal project management practice. They include the preparation of key specifications, estimating the project costs and time-scale, the secondment of key people to the project, sorting out lines of communication, organizing the process of formal project authorization, etc. Only those aspects which are important to new technology projects are developed here. The reading list for this chapter provides a broader range of project management topics.

8.3.1 The research proposition

When a research opportunity is assessed at the initial, tentative stage it is first related to corporate objectives and strategy. Is it compatible with these and does it foster them? If not, does the first promise warrant a change of strategy? If the opportunity is accepted it is then the task of the research and development function to develop first ideas into concrete proposals and plans of action and to submit these to the board of the company for authorization. In essence, this amounts to a set of benefit expectations and a measure of time and cost. It includes as accurate a forecast of development costs as is possible at this stage. Experienced research and development staff are likely to be more aware of the risks embedded in a project and what cost implications these have. Nevertheless, they must be able to defend their estimates against critical questioning. Of course, they normally draw on various functional departments for estimating support.

The biggest challenge with the R&D proposition consists of the various assumptions which underly it. They need to be made explicit and withstand detailed scrutiny.

8.3.2 Market expectations

Where the proposition involves the launch of a new product it should be axiomatic that its market promise is assessed first. This is not just the situation at the time of the initial evaluation, but what it is likely to be when the resulting product enters the market. At the overall market level, attention has to be given to trends. If the product is likely to be available commercially in two or three years time, thought must be given to likely changes in technology during such a lead time. This has been a particular challenge with the applications of microelectronics.

The new product must be seen in terms of the added value to the anticipated customer. Who precisely that would be, cannot be left to assumption. Especially a consumer product must therefore be specified in functional and in psychic terms; what appeals to the customer, the portfolio of his needs and the likelihood of the product meeting these needs. This involves the setting of a provisional target price which will be included in a product market specification.

The next step is the development of the product objectives:

- how it will be used,
- who will use it and maintain it,
- how it will be stored and transported,
- what the required performance and reliability will be,
- the conditions under which all these requirements have to be met
- the cost limits to achieve this.

In essence, the product objectives will set the boundaries within which engineering and design solutions have to be found. The product technical specification follows. This describes in detail the functions and the chief characteristics of the product.

8.3.3. R&D proposal preparation

The formulation of R&D proposals has much in common with the preparation of other technical proposals which are essentially requests to the company board for the authorization of expenditure. This section will focus on the aspects specific to R&D. General aspects of economic evaluation are described in Chapter 9.

The following aspects need to be considered and formulated:

- the precise objectives of the project;
- the technical relationships of the project with previous work;
- the technical feasibility of the project in terms of the available resources;
- the anticipated outputs of the project;
- the extent of competing R&D work in this field;
- the potential of the resulting products;
- the nature and extent of the anticipated markets;
- the extent of likely patent cover;
- the time-scale required to achieve the set objective.

An interesting comment about markets was made recently by a senior research manager in the chemical industry, who stated that the sales forecasts, obtained independently of the research workers concerned, had shown present values and returns that seemed extraordinarily high. In his view this was partly because there was an inadequate allowance for the possibility of research failures.

Initial sales forecasts are usually crude but if, in due course, the actual sales figures are fed back they provide a learning opportunity for research management. The more basic the research project and the longer the time-scale, the less predictable is the outcome.

8.3.4 Project evaluation

Because of the diversity of circumstances there is a basic need for a tangible decision reference system. The choice is seldom whether to carry out a R&D project or do nothing. When management has to decide how the company's R&D resources are to be used then the assessment will have to be in terms of the benefits, costs and opportunity costs of the chosen project. The opportunity costs are the benefits of the alternatives, the projects foregone because of limited resources.

Because of the nature of R&D and the associated uncertainties some companies are reluctant to carry out a full evaluation of R&D projects. Yet without an appropriate evaluation no meaningful scenarios can be established and no proper decisions can be made about the amount to spend on R&D. The evaluation of a proposition means that it has to be thought through, objectives need to be defined and their achievement transcribed into financial terms. Project schedules have to be linked to the time value of money.

There is the view that many of the benefits of R&D expenditure are intangible or at best very difficult to measure. While not denying the challenge of quantification it is argued that if a benefit is stated, it can be defined and if defined, it can be evaluated. It then becomes a matter of the degree of accuracy required.

The assessment of project worth has a number of aspects. Broadly speaking these can be divided into economic and non-economic facets. Even then the latter category has economic implications. The following are the more important categories of evaluation:

1. rating systems
2. risk analysis
3. financial evaluation
4. combined rating and financial analysis.

The evaluation can be in terms of factor profiles for marketing, commercial, technical, production and financial aspects. It can be numerical or on a judgement scale from excellent (E) to poor (P). Such systems normally use standard worksheets for analysis and subsequent project comparisons. Each major functional area has a set of detail questions to be considered by the specialist staff concerned. The manager responsible for the function carries responsibility for the contribution made by his staff. Such an overall framework thus collates the judgements of all responsible managers. Table 8.1 shows sample work sheets for a graphical display and numerical rating. The numerical rating can be further developed to include factor weighting.

8.4 RISKS WITH NEW TECHNOLOGY PROJECTS

You cannot plan success, but you can plan the activities which lead to it. One characteristic of the development of a new technology is the nature of the risks involved. Work may be carried out to schedule, but results cannot be guaranteed. If a prototype fails, then a new start has to be made. It is therefore important that the project systems are sufficiently flexible so that a development setback can be absorbed and revised schedules can be formulated without too much difficulty.

One risk of a R&D project is that it does not achieve its objectives. Typically, the technology, product or process may not work; the project specification has not been met or there has been a serious time or cost overrun. It is important to have contingency plans for such eventualities

Table 8.1 Sample illustrations of worksheet patterns

A – Technical factors	Graphical rating				
1. Utilization of company skills	E	Ⓖ	A	BA	P
2. Availability of staff	E	Ⓖ	A	BA	P
3. Availability of facilities	Ⓔ	G	A	BA	P
4. Time-scale for development	R	G	Ⓐ	BA	P
5. Scope for further projects	E	G	A	BA	Ⓟ
6. Scope for patents	E	G	A	BA	Ⓟ

Work sheet prepared by:
Work sheet approved by: Technical Manager
   E=excellent, G=good, A=average, BA=below average, P=poor

B – Technical factors	Numerical rating				
1. Utilization of company skills	2	2	3	④	5
2. Availability of staff	1	2	3	4	⑤
3. Availability of facilities	1	2	3	4	⑤
4. Time-scale for development	1	2	③	4	5
5. Scope for further projects	1	②	3	4	5
6. Scope for patents	1	②	3	4	5

Overall value for technical factors   3.5
Work sheet prepared by:
Work sheet approved by: Technical Manager
   1=poor, 2=below average, 3=average, 4=good, 5=excellent

when their likelihood emerges. Such fall-back alternatives should be assessed in the same manner as the original project proposals.

In the assessment of a research proposal a number of risks have to be considered and their implications assessed. The following are the most important:

1. unforeseen technical problems; the technical specification cannot be met,
2. R&D costs get out of control,
3. the resultant product/process is not competitive,
4. a competitor wins by securing key patents or winning the market,
5. the market disappoints – no product volume,
6. 'assumptions' have changed, such as with major shifts of raw material prices or exchange rates,
7. there are manufacturing problems,
8. product liability claims.

The systematic review of risks is not just to become aware of them but it is also the basis for risk reduction. The various risks of a project can be consolidated into an overall project risk index. Table 8.2 and Fig. 8.1 illustrate the concept. Essentially risk assessment has two dimensions here. The first is the probability of failure. The higher the chance of this, the greater is the risk weighting. It will be noted that the probability estimate does not exceed an even chance; in a commercial context a firm would be foolhardy to embark on a project if failure is more likely. The other dimension is the stage of assessment. For instance, if a problem is unresolved at a comparative late stage of development, the risk of that problem remaining unresolved will be correspondingly greater.

Table 8.2 The concept of project risk rating

	Risk levels	
Risk evaluation	*Chance of failure (%)*	*Weighting*
None	0	0
Small	1–10	1
Some	11–20	2
Substantial	21 30	3
Major	31–50	4

	Levels of investigation	
Level	*Development*	*Marketing*
1	No investigation	No investigation
2	Feasibility study	Market assessment
3	Prototype stage	Market research
4	Full design stage	Product specification
5	Pre-production	Launch stage

Figure 8.1 shows a sample risk rating for a rotary compressor. The risk index is the product of the risk rating and the investigation weighting. It is helpful to have separate risk indexes for critical and contributory project features. A feature becomes critical when unsolved problems mean basic technical or market failures. A contributory feature will lead to the impairment of project prospects, typically by causing schedule and cost overruns.

Basically, the analyst judges the risk on a point scale and relates this to the degree of work already done to eliminate it. The bigger the risk and the longer the related problems remain, the higher will be the risk index.

While such a technique has a number of limitations, such as the subjective nature of risk assessment, it also has some benefits. First of all, it provides a systematic approach for different staff to use. It can, therefore, be used for discussions or brain storming sessions. Secondly, it provides a

Project feature	Nature of risk	Risk rating x = 0 to 4	Level of investigation y = 1 to 5	Risk index x · y
A – Critical features				
Vibration	Prototype trials troublesome	4	3	12
Alloy castings	Samples from suppliers defective	2	2	4
Others				
Total				16
B – Contributory features				
Manufacture	Assembly methods unresolved	2	1	2
Design	Risk of patent action	1	4	4
Others				
Total				6

Figure 8.1 Sample risk rating work sheet.

more structured form of risk comparison between different project options. Thirdly, it furnishes a time series for a risk review at various project stages and allows more informed stop loss decisions to be made. Milestones can be defined for risk reduction.

Although the assessment is subjective, it is often better than 'intuition' which often remains a mystery. At least some of its judgement can be developed by case experience and guidelines can be prepared. Any pattern of values can be used, provided they are applied consistently.

Parallel development efforts are desirable where there are high initial technological risks, provided these can be reduced within a contained time/cost framework. Otherwise there will be heavy escalation of R&D expenditure.

8.4.1 The assessment of technical success

Related to risk assessment, and a crucial aspect of project evaluation, is the degree of likelihood that a project will be a technical success. It is important, therefore, that a clear specification is established as to what is required to demonstrate technical success. Such an assessment requires the tabulation of a set of succinct statements which can be encapsulated in a standard work sheet. Figure 8.2 is an example, applied to an automatic assembly machine for electronic components. Here a proposed equipment specification is related to its prospects of success. For instance, if the proposed assembly rate is well below existing practice, the prospect will obtain a full rating, as any associated problems seem already resolved. Where an entirely new attribute is specified, such as the type of feed system, then the prospect of success is downgraded.

Factor	Value	Excellent 100%	Good 75%	Average 50%	Below average 25%	Poor 0%
Speed	40	Well below existing practice	Same as those in use	Faster but attainable	Faster but solution possible	New solution needed
Versatility	15	Completely versatile	All existing uses possible	Most uses possible	Some uses possible	New solution needed
Board size	15	Takes all possible sizes	Takes existing sizes	Some boards need changing	Takes some boards	New solution needed
Feed system	30	Takes all likely components	Takes existing components	Some changes needed	Takes some components	New solution needed
Score maximum actual	100 55	40		7.5 + 7.5		0

Figure 8.2 Sample technical success assessment.

8.5 PROJECT ORGANIZATION

Whatever the form of project organization it is important, in the first instance, to have a 'project champion', who normally is a board member or senior manager, committed to the successful outcome of the project. Often, the project or programme manager reports directly to him. He can clear opposition to the project and defend it when there are competitive pressures for the allocation of resources. Particularly with a product embodying the new technology there is a great need for close links between design, manufacture and marketing. The integration of these functions to meet project needs is a particular task for such a champion.

The following are the main forms of project organization. The specific form is a function of the scale and circumstances of the project.

8.5.1 The programme management approach

A programme manager is responsible for the carrying out of a R&D or a DDMM project. His responsibility is the direction and integration of the required work carried out by different functional departments. As the programme manager has no line authority to instruct specialist staff, much of the success of his role depends on his personal skills. Successful interaction with colleagues, particularly as part of project coordination, is crucial. The seniority, status and capability of the programme manager is thus an important contributor to the speed and the outcome of the project.

An example of such an organization is provided by the new model development system operated by Volvo, the Swedish car manufacture. Volvo have programme managers, with director level status, who are responsible for leading development projects right through from early concept to product launch. Within each function, such as manufacture, responsibility for a development project is given to one person who is responsible for clear decision making.

8.5.2 Matrix management

With this form of organization specialist staff are seconded to the project and are responsible to the project manager for the effectiveness of their project contribution. What needs to be achieved is expressed in target schedule and cost terms. For their professional skills they remain responsible to their functional heads of department who are responsible for how these targets are to be attained. The simultaneous responsibility to two managers has the risk of potential conflicts and dilemmas for the specialists and it is important for the project manager to emphasize procedures to avoid this. Specialist staff may be seconded to the project on a full-time or part-time basis.

There are two basic approaches to matrix organization. The first is known as the leadership or project matrix. The project leader is expected to motivate his team to work towards project goals, needs status and leadership skills. He has direct authority to make decisions about work flow and the disposition of staff. The second is a coordination matrix, where the leader is more of an information centre and a scheduler of required contributions.

8.5.3 Task forces

These consist of an interfunctional group of staff, often brought together at short notice, to deal with particular assignments or programme bottlenecks. Their activities are usually intense but of relatively short duration. They need to have clearly set goals and the backup of a senior manager who is the task force champion. In some companies, task forces also serve as a management development opportunity for young staff who are thrown in 'at the deep end'. Much of their relationship is informal.

8.5.4 Dedicated teams

Functional specialists are brought together to work full time in a separate location on a particular project. The nature of the work is such that the various specialists will have to widen their range of activities to contribute to the progress of the project. This pattern of organization encourages transfer of know-how between different functional specialists. For instance, Toyota organize their development projects around a **Susha**, an energetic, single-minded individual who takes total responsibility for the project from start to finish.

8.5.5 Project systems

Particularly for new technology development, project systems need to simplify problem solving across functional boundaries and ease information flow. Informal communications are encouraged, but behind them structure is required. Many companies which have a number of projects at any time have manuals of procedures. A modular presentation gives flexibility for varied project circumstances. Routine activities are standardized as a basic support for flexible operations, so that creative time is better used. Information is tailored to assist the decision requirements for responsible managers. The most relevant systems are for project planning, reporting and cost management. These will be dealt within subsequent sections.

8.5.6 The use of consultants

While most large companies carry out their own R&D projects, the limited staff resources of smaller firms may require them to seek consultants in addition to their own staff. This also pertains to the introduction of new technology from outside, such as automation schemes where advice and guidance on implementation schemes can be critical to the success of the innovation. A comparative situation can also exist where a business unit within a large firm uses, and is charged for, corporate specialists. The basic principles of project management apply just as much to these cases as to internal company projects. The company, which is the client in contract, usually appoints a responsible manager for the project, with or without the formal title of project manager.

Bessant and Grunt (1985), in their comparative study of management and manufacturing innovation in the UK and the then West Germany, specify some important aspects in the relationships between company client and consultant. They stress the importance of clear terms of reference, proper specifications and documents, so that both parties know the overall intent and the specific objectives of the assignment. This is important at the start and for the duration of the project to ensure continuity in case of personnel movements. There need to be explicit statements about the location of all the various work stages and the contribution expected from client staff. Liaison procedures must be agreed and liaison staff appointed. This becomes crucial when there is a significant development/application engineering content in the project and changes are required. With a development project, security aspects can become an important contract feature. Above all, there is the general need for agreed reporting methods, progress meetings and all the other administrative requisites of good project management.

8.6 PROJECT PLANNING

Because of the uncertainties affecting R&D projects, where some of the main task stages are hostage to the success of and the data derived from prior work, some of the planning has to be done in stages. Nevertheless, there is a need for planning discipline where the initial stages can be defined and when subsequent courses of action can be determined. This applies particularly to the following:

1. The definition of the main work tasks and their subdivision into appropriate hierarchies and work packages. The more specific and detailed these can be made, the better is the chance of success, but because of the inherent uncertainties of a development project it will be difficult to specify these in full at the outset. While the technical requirements of a work package may be established, its task content could be hard to assess. In such cases task durations are often estimated on a pessimistic, optimistic and most likely basis.
2. The development of schedules, charts, precedent diagrams and networks. Networks are particularly important to identify the critical activities and path which determine the progress of the total project. The most common are CPA (critical path analysis) and PERT (programme evaluation and review technique). The techniques are readily available in standard software package form and can be expressed in the pessimistic, optimistic and most likely format. More advanced forms of statistical analysis can be applied using probability values for different scenarios. As well as activities the schedules can include key decision calendars.
3. The marshalling and scheduling of resources to meet the defined timetable. This includes all functional staff as well as key equipment, such as the use of electron microscopes.
4. The definition of milestones. These correspond to the key stages of the project. A milestone matches required results with target dates; their aggregation yields the project target date. For example, by 30

September a team may have to complete a design, validate a method, set up a pilot plant, complete a test programme, obtain regulatory approvals, such as a certificate of airworthiness. It is important to obtain team commitment to milestones at the beginning of the project.

5. The specification of a project cost structure and accounts plan. One of the great risks with development work is a cost overrun. The emphasis must therefore be on prompt cost reporting and on cash flow control. Staff costs accumulate relentlessly and project documentation must highlight the use of staff time.

6. The formulation of project control systems. Apart from cost control these will relate to progress against schedule and technological achievement.

8.7 PROJECT REPORTING

It is a common feature of successful development projects that the project manager is close to key colleagues and their activities. Much of the project reporting comes to him in an informal manner. Similarly, the project champion and other top management may be well supplied with informal information. Such a setting is useful but in no way suffices, a formal procedure is essential.

Most large companies carry out a number of development projects and have standard reporting procedures. It is common to integrate these into manuals, particularly the financial reporting requirements.

A reporting system has to meet the following objectives:

* to report progress to top management;
* to inform all concerned working on and alongside the project;
* to concentrate minds on action by exception reports.

Reports are best structured on an hierarchical basis and made appropriate to the decision makers who have to act on their content. There is the temptation with computer-based systems to overload top management with detail. Briefly, what matters is the following:

* to state the current position;
* to compare it with the target position;
* to show the variation from plans and variances in costs;
* to explain the reasons for the differences;
* to forecast the implications of the deviations;
* to assess the chances of technical success.

Procedures need to highlight decision stages, milestones and a timely feedback of results. It is particularly important with R&D projects to have interval estimates of technological success. These can be given, typically, in probability, schedule and cost terms. Such management information is crucial for the early detection of adverse trends. Simple, well-defined systems, clear and readily understood by the staff concerned need to be in place. It is also important to have unobstructed and prompt information flow across functions. Documents should be simple but be designed to highlight key control information. A wide range of computer software packages now facilitate this.

Project reporting is usually on a period basis. For instance, a monthly reporting procedure fits in with most financial control systems. However, reporting by milestones, which are clearly ascertainable stages in a project, gives a better relation between costs to date and progress achieved. The actual reporting document could be a standard form with graphs and appendices as required, a computer printout, or part of an electronic mail system. Figure 8.3 is an illustrative example of a project

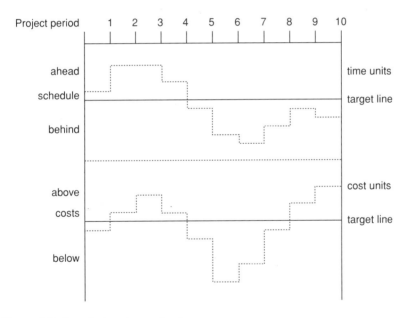

Figure 8.3 A project variance chart.

presentation which concentrates on the deviations from plans. The horizontal lines are the targets for each project period; deviations are shown by the dotted lines. For instance, by the end of project period 3, the project was ahead but also showed corresponding expenditures.The end of period 9 indicates that efforts were made to catch up with the schedule but costs have exceeded the budget rate.

Another useful graphic presentation is a 'slippage' chart, which highlights the overdue activities of a PERT (programme evaluation and review technique) chart.

8.8 PROJECT COST MANAGEMENT

8.8.1 Cost and cash flow aspects

For many companies the escalation of development expenditures is a major problem and the resultant drain of cash can be disastrous. Technical excellence is no substitute for financial discipline and the cutting edge of innovation is not necessarily a path to riches. Effective project cost management is a basic need for excess cost avoidance (see also Chapter 9). Wherever possible, managers should check on the cost outcomes of previous comparative development projects and ascertain the reasons if there were cost overruns. Where some of the development or design tasks are contracted out, tight cost control also tends to tighten the work schedule as contractors are anxious to avoid non reimbursable costs.

Companies which undertake a number of research projects usually have a standard accounting and cost control practice. Special systems may have to be installed where a research project is part of a consortium arrangement or a contributory funding body has special requirements.

Cost management starts with the estimate of the costs of a proposed research project. The accounting framework for the estimate usually matches that for the project execution. A strong argument for the association of the prospective project manager with the estimate is the continuity of judgement. The manager who requested funds from the board makes a commitment to carry out the authorized project.

The following are the main requirements for effective cost management:

1. The use of an hierarchical accounts plan. An accounts structure which can cost and aggregate work packages is essential. Without it, estimating accuracy is impaired and the subsequent allocation of responsibilities is diluted. Most companies use standard accounts codes for ease of estimating and the subsequent posting of expenditures.
2. The establishment of cost centres. These are requisites for the allocation of funds and the appraisal of performance of major activities, such as electrical design, the model shop etc.
3. The preparation of budgets for the cost centres.

4. The prompt reporting of expenditures. With a large project this is often the responsibility of a project accountant.
5. A well-established period/milestone reporting system. This is best achieved as part of the overall project reporting system. It should give the period costs, costs to date, variances and forecast completion costs for each cost centre. Where the work in a cost centre has a large routine component, such as detail design or testing, an index which relates the value of work done to a *pro rata* budget gives a better picture of progress achieved.

8.8.2 Cash flow curves

One of the most telling correlations for many development projects is that of a late project which has overspent. The longer activities take, the more time bookings will be against the project. Setbacks and delays affect the project cashflow. The following figures show the implications. Figure 8.4 shows the effect of a project schedule overrun where costs have still been kept within budget. The immediate result of delay is the postponement of cash inflow. The payback period is correspondingly extended and will take a larger fraction of the project life cycle. If the project is financed by loans, the project interest charges will increase accordingly.

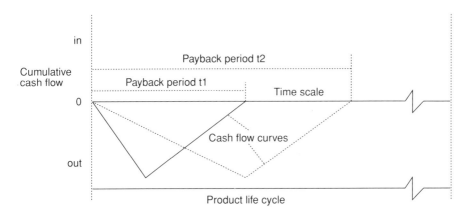

Figure 8.4 The cash flow implications of time overrun.

Much more likely and even worse will be a schedule and cost overrun. This is shown in Fig. 8.5. The further drain of cash will extend the payback period. Other measures of project evaluation will show corresponding deterioration.

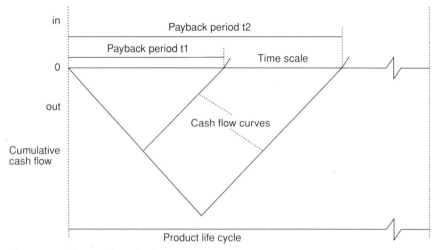

Figure 8.5 The combined effects of schedule and cost overruns.

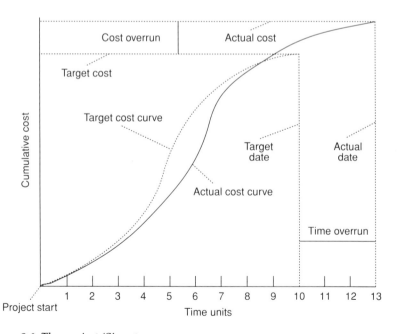

Figure 8.6 The project 'S' cost curve

8.8.3 The management of the development 'S' curve

Figures 8.4 and 8.5 show the impact on project cash flow with its implications for project profitability. To anticipate and possibly to avoid such unfavourable outcomes, it is desirable to track cashflows with an 'S' curve, as illustrated in Fig. 8.6. The initial 'conceptual stage' will be the relatively flat part of the curve.This pre-development stage is essentially concerned with feasibility confirmation and risk reduction. The curve will come to the steep middle section when the project R&D expenditure reaches the full development stage and a larger number of staff are involved in detail design or testing.

Where a company has the documented cost experience of a number of development projects it will be able to use 'S curve analysis' to compute the likely cost and schedule outcome of a specific project when the cost data for, say, 25% of the overall project period becomes available. In addition to cost control such curve analysis can assist with strategic decisions. For instance, the pacing of development work can in some cases be matched to suit market opportunities.

8.9 THE CRUCIAL TIME ELEMENT

Give a development project half a chance and it will slip! With the shortening of product life cycles the time during which a product and its technology can make a contribution to the revenues of the company will be significantly diminished. This reduction will be disproportionately greater if the lead time for product launch remains unaltered. Also, as the pressure to reduce such life cycles is due to competition, the risks of being overtaken are greater. This makes the reduction of lead time a vital requirement for project management. The emergence of such techniques as concurrent engineering or rapid prototyping, as mentioned in Chapter 6, are responses to this need; there is an even greater premium on time conscious project management. The value of time is a continuous aspect of trade-off decisions. The delay with an 'optimum' solution cannot be justified when an 'adequate' solution gives a time advantage.

Figure 8.7 shows a product life cycle in company cash flow terms. The product life is the period between the product launch and the product cut-off. It will be noted that the product resource commitment time span is larger than the product life and that the cash outflows are at the beginning.

Where a product is subject to sudden competitive threats its life cycle pattern is approximated by Fig. 8.8. In this diagram:

t = 'traditional' product life – customer determined
b = 'turbulent' product life – competitor determined
a = product development period.

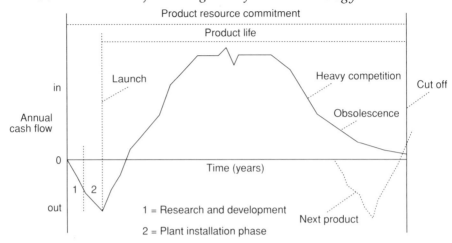

Figure 8.7 Product life in cash flow terms.

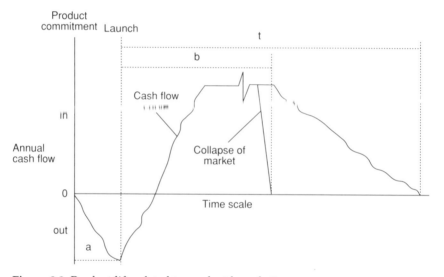

Figure 8.8 Product life related to product launch time.

There will be no gentle decline of a mature product; the drop will be rapid, orders will dry up. The commercial success of the product is very much a function of the ratio 'b/a' and the project manager's contribution is the minimization of 'a'.

8.10 THE REQUISITES FOR CONTROL

Despite the informality of participative working there is the need to achieve and to attain agreed objectives. The successive stages of a development project call for different emphasis. At the most creative stage, when solutions to key problems begin to emerge, too tight a rein may cut off other likely solutions and cause frustration which will limit creativity. But when projects mature and the path is well defined, designs need to be frozen and target dates held.

It is the task of any reporting system to furnish the basis for effective control. The frequency, manner and the extent of documentation are normally designed to achieve this. It is, nevertheless, important for some project aspects to be emphasized for more effective control. These include the following:

1. Project definition. A clear statement is required of the project objectives. These need to include the criteria for project success, expressed in terms of results, costs and time scale.
2. Updated expectations. At each reporting stage a revised statement of anticipated completion costs and expected completion dates will be required.
3. Staff deployment summaries. As staff inputs and costs are critical with new technology projects these must be related to project progress.
4. Technical feasibility status. When the assessment of technical success indicates sufficient promise to proceed with the project, the steps towards its realization can be expressed in hierarchical form and linked to corresponding time and expenditure schedules. This may then be related to commercial requirements, such as a critical market entry date, e.g. an important trade exhibition.

The main technical objectives can be translated into project milestones and progress is denoted by the technical feasibility status which expresses the optimum/pessimism of the project manager as to the likely outcome

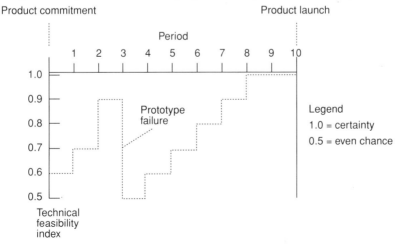

Figure 8.9 A technical feasibility status graph.

of the project. For instance, one multinational corporation starts its projects at 0.5, i.e. there must be at least an even chance of the technical success of the project. If there is trouble, the index will decrease toward 0.0; when the problems are resolved the index approaches and attains 1.0. A graph of this can become quite telling over a period. Figure 8.9 shows the use of such an index.

The index also allows management to re-assess the minimum and most likely activity times required to achieve the stated technology requirement. Sometimes too, the earliest activity time needs to be established at which failure has to be acknowledged. This is particularly important where project funding is by stages and depends on the progress made.

5. The rate determining factor. This is the factor which controlled the progress of the project over the last review period. It is akin to the critical path concept in network analysis.

8.11 THE COMPLETION OF NEW TECHNOLOGY PROJECTS

A project which has achieved its technical, budget and schedule objectives will come to a natural close and can be wound up in a standard manner established by the company. Frequently, a formal notice is given to all those linked with the project, together with a checklist for the winding up process. The retention of files, computer files, documents and other data should reflect their learning value for future projects. It is also important to stop time bookings against the project. This is normally done by the closing of an accounts code. However, many projects have residual requirements, such as may arise from a product launch. It is a matter of judgement whether these should be hived off and the main project closed. It is appropriate, but not always comfortable, to carry out a project audit when the project comes to an end. This final stage is treated in Chapter 9.

8.12 THE CLOSURE OF NEW TECHNOLOGY PROJECTS

More of a problem is the project in trouble. A company may have to face the unpalatable and close a project. With an effective reporting system some project difficulties can become apparent at an earlier stage and give the company an opportunity for review and rearrangement. But there are also cases where an unexpected event such as a new law or a catastrophe elsewhere requires a prompt decision. It is therefore important to have a formal procedure for winding up and terminating projects. An aborted project means unrecoverable expenditure and requires explanations.

Before leaving this chapter it is pertinent to report some reasons for project failure. Mansfield (1969) in his study of the R&D projects of a major US electronics firm found the following causes to be significant:

1. manpower diverted to other projects,
2. change in project objectives,
3. key staff leaving,
4. technical problems.

From a project management point of view it is interesting that three out of the four main reasons were managerial, not technological.

8.13 SUMMARY

Project management for new technology has to face the uncertainties of research and the challenge of cost escalation with development work. The formulation of R&D proposals requires the careful integration of different functional analyses and the systematic application of established evaluation and risk assessment techniques.

Various forms of project organization can be selected, depending on the scale, urgency and complexity of the project. Project planning has to allow for the variability of work package task contents and project reports need to focus also on the probabilities of technical success and the work implications if it has not been realised within the set time-scale. The containment of project costs and the adherence to schedule warrant continuous attention.

FURTHER READING

Harrison, F.L. (1992) *Advanced Project Management*, 3rd edn, Gower Publishing Co Ltd, Aldershot.
Knight, K. (ed) (1977) *Matrix Management*, Gower Press, Aldershot.
Lock, Dennis, (1992) *Project Management*, Gower Publishing Co Ltd, Aldershot.

CHAPTER 9

Investment in new technology

9.1 INTRODUCTION

Kodama (1991), in his telling analysis of Japanese high technologies, indicates that research investment decisions by Japanese companies are no longer based on rates of return. As he puts it bluntly: you invest or you die. At the other end of the spectrum, many Western companies have demonstrated that you can invest and still die or even that you die because you have invested.

Commitment to a new technology, just as any other commitment, requires an economic assessment. This chapter is concerned with how this is carried out: the available techniques, the role of judgement; the caveats and the benefits that can flow from it. The company which is prepared to commit itself to substantial expenditures wants to reap corresponding benefits from these. It wishes to assess, as far as it can, the worth of such investments.

There are two major forms of investment in new technology. The first is investment in the development of new products and processes. Usually the bulk of such expenditure is within the company. The second is the acquisition of equipment and systems which embody the new technology. The main expenditures here are those of procurement.

The focus of this chapter is on major investments which have a significant effect on the competitive position of the company. Of course, in a period of rapid technological change a company may acquire new technologies embodied in equipment as part of routine plant replacement operations. But such purchases are not necessarily strategic.

An investment is defined as the acquisition of an asset which yields future incomes. Because of this the asset will have value. The discounted value of the resultant income is expected to be greater than the cost of the asset. The pursuit of knowledge which is the hallmark of 'pure research' will not be included here. As will be discussed in Chapter 10, such expenditures are written off against the profit and loss account. Both, applied research, which yields patented products or process applications, and development, which transforms an invention into an innovation, can be considered as investments.

There are, of course, many instances where a firm carries out such investments as part of a major modernization project. For instance, the machinery division of a UK manufacturer of plain bearings carried out a major development programme which transformed conventional production engineering equipment into an integrated automated production line. The equipment as it was developed, designed and built was then priced to include appropriate recovery rates for development costs. The operating divisions, which were separate profit centres, then acquired these units as if they came from external suppliers. From a corporate point of view both the development and the plant investments were part of its overall strategy. The newly developed machining lines were also sold commercially, with a substantial proportion of exports.

9.2 INVESTMENT AS AN IMPLEMENTATION OF STRATEGY

Investment here focuses on the major projects which express the business strategy of the firm. It is the integration of strategy and economic assessment which is the basis for investment in new technology.

The strategic evaluation of such a project consists of the following:

- investment evaluation
- scenario cost/benefit analysis.

No major investment is meaningful without an economic analysis and, bearing in mind the fragile nature of some of the forward projections in an investment calculation, there also needs to be the evaluation of strategic and intangible aspects that, for the want of reasonably accurate financial information, might be overlooked.

The investment in new technology, particularly production processes and equipment, can make heavy calls on the funds available to a company. Those calls compete with other claimants for funds, such as the development of new markets and products. Technology, as a claimant, is not sacrosanct and a company has to ensure that the opportunity costs of projects foregone are less than the likely benefits promised by technology investment proposals.

9.3 INVESTMENT IN R&D PROJECTS

Surveys of R&D expenditure by US industry estimate that 87% of the expenditure was devoted to new products and the improvement of existing products. The remaining 13% were spent on new processes. With such a high proportion of expenditure devoted to products, the risks of the market place in which these were sold were immediately relevant. With many companies market risks were regarded as higher than those involved in attaining technical project completion. Within the chemicals and drug industries a study by Mansfield *et al.* (1972) on the termination of R&D projects, before their technological objective had been achieved, found that about 60% of projects were abandoned because of commercial reasons, such as inadequate margins or markets, while 40% were technically unsuccessful.

The basic evaluation requirements for a R&D proposal are typically:

1. the estimated cost of the R&D programme,
2. the time series of R&D expenditure,
3. the probability of technical success,
4. the probability of commercial success,
5. the extra sales value generated,
6. the generated extra income time series,
7. the rate of return forecast for the project.

A major, but not the sole 'success criterion' for a project, would be a rate of return greater than that available from an investment elsewhere.

It will be appreciated that the financial worth of a project proposal depends on the accuracy of the cost, development time and income predictions. It is very difficult to be accurate with a long-term forecasting period and it is important that the rate of return prospect is related to other technological and business factors. Also, uncertainty about the outcome of the major project parameters can remain well into a project period.

9.3.1 Investment in product value added

For products already on the market, Edge (1990) emphasizes that competitive advantage in the market place is controlled by value added, particularly because of the pricing flexibility it furnishes. A successful investment in a particular R&D project will therefore generate an incremental added value to a particular product. This can then be multiplied by the expected product volume to give a time series of added value generated by the project. Such a time series is a measure of the worth of the investment.

Added value also can be in the form of good application engineering, yielding special value to a customer. The location of value added can be in various functional areas, such as more effective manufacture or special service facilities.

9.3.2 The interaction effects of R&D projects

Much has been written about the difficulties of forecasting the expenditures of development programmes. This is especially the case with multi-stage projects; where the cost of a later stage is a function of the success of an earlier stage. For example, a UK engineering company decided to introduce automation into the manufacture of thin wall bimetallic plain bearings for car engines. The first stage of the modernization programme was the development of a continuous sintering plant to form a range of metal strips from metallic powders. This had to deal with problems of sintering, casting, continuous operation and plant design. Then followed the rolling process to secure proper bonding of the metallic constituents and the press operations which cut and shaped the strip into semicircular bearings. Again, this involved metal forming and metallurgical problems. The total manufacturing process which included some subsequent feature machining was to be integrated into an automated production line. The range of development problems to be solved was diverse, interdependent and required different technical resources. There were great difficulties in forecasting the total cost of the project although technical feasibility was recognized. The company ultimately decided on a step-by-step set of projects where the success of one left the options open for the next stage.

9.4 ESTIMATING FOR DEVELOPMENT PROJECTS

Most companies which carry out a significant amount of research and development have standard accounting systems for the estimating and control of development projects. Batty (1988) has provided a good treatment of the subject.

9.4.1 The definition and content of development tasks

The first step in the estimating process is to determine and to structure the main development tasks. This is often already part of the feasibility analysis.

The content is then expressed in terms of the expected activity levels of staff in the various technical cost centres. As an example, in the engineering industry this could typically involve: design (mechanical – electrical and electronic), software development, prototype work/model shop, test and reliability confirmation etc. Section heads would be responsible for estimating, in conjunction with estimating engineers, the expected work content in man/months or other similar units. This is crucial as with development projects staff time is generally the main cost function.

9.4.2 The establishment of development budgets

The following are the main constituents of a development budget:

1. Staff costs. The transcription of staff time into money is the basis for the preparation of the cost centre budgets. With many projects this will be a time series spanning over a number of project periods.
2. Equipment. This includes all the equipment required for the execution of the project. In certain cases there will be hire charges or other payments e.g. the use of time on a university electron microscope.
3. Consumables.
4. Escalation. Where a project is likely to span several years, a provision is generally made for increases in salaries and material costs.
5. Contingencies. In Chapter 8 we already noted the spectrum of risks associated with new technology projects. The great difficulty is how much to allow for contingencies where there are risks of schedule and cost overruns. When the desired results are elusive and project staff continue to work beyond the target dates, delay and excess costs escalate. Mansfield *et al.* (1972) in their studies of R&D project completion found that commercial projects typically took twice as long and cost twice as much as their original estimate.

9.5 FINANCIAL RISK REDUCTION

The possibility of cost overruns is a continuous preoccupation for technical management. Careful estimators and managers will critically scrutinize staff time budgets and question the assumptions on which they are based. Wherever possible they will use cost records from previous projects to guide them. Some of the techniques listed below provide guidance for their estimating tasks.

9.5.1 The use of improvement curves

It will be recalled that a number of tasks, particularly with development work, are relatively routine. Examples of this are detail design work; routine testing of drugs; component certification tests, etc. An improvement curve can be drawn to show the reduction in unit costs as the frequency of a task increases. In principle, this type of curve is similar to the learning and cost reduction curves used in batch and volume manufacture. One major improvement is the reduction in task time as experience suggests improvements which are incorporated in the task structure. Furthermore, equipment and tools become economic as the frequency of a task increases. An experienced industrial engineer can forecast such improvements in terms of task numbers and can estimate corresponding future costs. The technique was first applied to prototype/pre-production batches for aircraft. Figure 9.1 shows an example of an improvement curve in index form. The slope of the curve is determined by an 'improvement coefficient' which represents the rate of cost reduction. In turn, this is a function of the work task. The relationship can be expressed as follows:

$$C(n) = C(1). n^{p-1}$$

where $C(n) =$ unit cost of the 'nth' unit
$C(1) =$ unit cost of the first unit
$p =$ improvement coefficient
$n =$ quantity or frequency of task

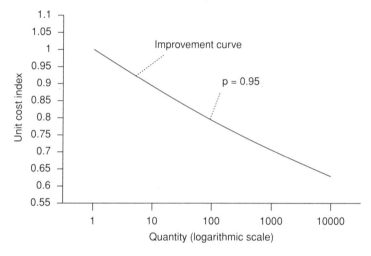

Figure 9.1 An improvement curve in index form.

9.5.2 Specification of the upper limit of financial risk

The increasing cost of advanced technology development and the current strong preference of government departments to award contracts on a fixed cost basis, makes the containment of development projects costs imperative. A good review by Andrews (1971) describes the nature of cost overruns and how these could be contained.

Basically, cost overruns are due to the following:

- unexpected increases in wage and salary costs,
- miscalculation of the programme costs,
- omissions in the estimate of much of the work that is required to achieve the proper objectives.

The avoidance of errors and the responsibility for specific provisions, such as for inflation, are relatively straightforward matters of management. The third category is the challenge and the best answer here is the rigorous definition of project requirements and engineering specifications.

Tischler (1969), in his assessment of the costs of space projects, noted a strong correlation between the observed mean cost escalation and the degree of novelty incorporated in a development. The greater the anticipated technological advance, the bigger is the uncertainty about its attainment which, in turn, is likely to generate more changes. The more changes, the greater is the cost escalation. Where changes lead to the 'recycling' of the development plan, costs usually increase more rapidly because of the duplication of previous work. To avoid this, it might be better to undertake the independent development and the retrospective fitment of novel devices, so as not to unsettle the main development programme. In any case, subsequent development after the main product launch could help to maintain the product's market position.

Tischler classifies changes in terms of the expected degree of technological risk. The terms used are semi-qualitative descriptors.

1. Modification. This is a change to an existing product or process which is known to work. The change may be a functional improvement or to suit a new application. The changes involve little technological risk and the resultant cost of the change can be accurately assessed
2. Interpolation. This covers, typically, new designs with features that have not been developed before but which are within a range of established knowledge. Development costs are within a range of likely figures, but their precise location may be difficult to estimate.
3. Extrapolation. The development work goes beyond established experience. Previous cost patterns may be useful indicators, but the degree of risk is considerable as there could be discontinuities in cost curves, e.g. with continuous miniaturization of components.
4. Innovation. Tischler uses this term when something completely new is undertaken and no experience is available to guide decisions and assess risks.

Figure 9.2 shows the cost curves which Tischler uses to illustrate his concepts. It will be noted that both the cost escalation factor and the project cost dispersion increase with technological risk.

It will be appreciated that such a basic classification is conceptual and oversimplified. For instance, even with an innovation project there will be tasks at the interpolation or even the modification levels, while many a modification project encounters extrapolation problems.

9.5.3 Excess cost avoidance (ECA)

The objectives of this method are to identify untoward events and to modify project schedules so as to minimize their impact. The ECA method concentrates on the worst probable combination of unfavourable

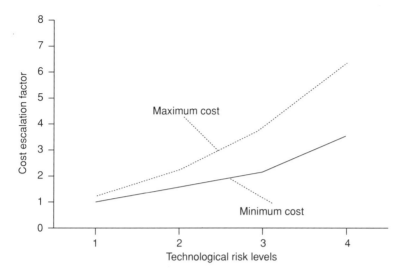

Figure 9.2 The cost implications of technological risks.

developments and, after systematic and consistent assessment, determines the likely limit of financial risk. It also considers the best probable combination of favourable events and so establishes a range of outcomes. As Andrews so aptly puts it, ECA explores what happens to a programme when every assumption is found to be invalid, every deduction wrong and every planned event frustrated or late. The investigator concentrates on the project network and assesses the implications when an event has not happened at all or not at the scheduled time. Obviously, the scope of mischief is infinite and in practice the investigator would concentrate on key events.

9.6 INVESTMENT IN NEW TECHNOLOGY PROCESSES

This is the second form of strategic investment and is mainly concerned with the introduction of external technology to manufacturing. Much of the investment costs will be the quotations of suppliers offering the new, embodied technology. Unless the proposed installation is the first of its kind, the risks of the new technology, *per se*, are reduced, but the risks of application remain.

9.6.1 Organization for project preparation

(a) Defined responsibility

A strategic proposition is best prepared without the operational pressures of day-to-day management. The responsibility for the preparation and submission of the investment proposal needs to be given to a project manager, a named senior staff, either a functional specialist or a line manager seconded to the project. Time is needed to think through the associated problems. Strategy as a rush job is the basis of disaster. Of course, as already noted in Chapter 8, a number of other staff will be involved in the detailed preparation of proposals.

(b) The steering group

The technological, financial and operational complexities of a major project require a dialogue between the project manager, his colleagues and members of the company board. In the first instance, a steering group, consisting of senior managers, is a source of authority and guidance for the project manager who, with his staff, will need to gather a wide range of information from different functions within the firm. Secondly, the dialogue with the steering group will be a major channel of information both ways. The project champion will be a key member of the steering group.

(c) Consultants for small companies

Where staff resources are limited and capital expenditure budgets are small, substantial plant investments tend to be intermittent and the expe rience gained from one project could well have been lost before the onset of the next. That is quite apart from the technological advances incorporated in the next generation of equipment. Where the rate of technological change, embodied in new equipment, is modest, such limited experience is not an unsuperable obstacle, but where there are rapid advances, such as with manufacturing automation, it is a major step. Often the company lacks experienced technologists who can effectively pick up and properly evaluate automation opportunities. A consultant can provide the expertise, but, in turn, needs to be chosen with care.

9.7 ASPECTS OF EVALUATION

With major plant installations the evaluation process becomes more complex and there is a tendency for various groups to concentrate on different parts of the analysis. The manufacturing engineer, typically, calculates and extrapolates from the unit savings achieved by the new process and plant. The accountant attends to profit contributions and tax implications. While such operational division of labour is understandable it often results in the fragmentation of analysis. One common problem is the duplication of effort because of an uncoordinated approach. The other is the unawareness of critical assumptions which are not unearthed by a full dialogue. The optimism of the manufacturing engineer may not be shared by his financial director.

An integrated approach to the economic assessment of new technology could avoid such a dichotomy of judgement. Four stages of evaluation can be discerned here, from the local and specific analysis to the overall computation. Table 9.1 shows the main constituents of this approach.

The summed effects of one stage become a constituent of the next. It does not, however, cover every feature that needs to be taken into account by a particular manufacturing company; rather it provides the framework for a more comprehensive and specific check list. A good example of how such a systematic approach can be developed is given by Warnecke and Schraft (1982) in their work place analysis for engineering production with particular reference to robots.

Table 9.1 The hierarchy of economic assessment

Stage	1	2	3	4
Level →	The production task	The production quantity	Aggregate production	Financial evaluation
	Task configuration	Pre-production costs	Work mix	Capital cost
	Speeds and feeds	Task volume	Utilization levels	Annual benefits
	Cycle times	Repetition factor	Downtime losses	Allowances
	Production rates	Tooling	Maintenance expense	Tax rates
	Labour content	Accessory costs	Variable overheads	Target returns
		Inventory/ WIP costs	Gain aggregation	

9.7.1 The production task

The analysis of methods and task elements reflects basic industrial engineering practice and contains both operational and handling operations. It does not follow that every task element has to be automated, only those which yield the desired return. That does not mean that the non-automated task elements remain untouched. In one example of robotic application of adhesive beads to car body parts the loading, clamping and unloading tasks remained manual but were combined with a press operation. The adhesive beads, however, were placed by a Unimate PUMA robot. The matching of cycle times between different work stations was important; it was the balanced line output, rather than individual machine cycles that gave the benefit.

9.7.2 The production quantity

For an automation system, such as a robot installation or CNC machining centre, the main preparation costs are programming and application proving. While these are considerable for the inexperienced firm, there is a major learning curve benefit after such a system has been in operation for some time. For instance, the use of pendants by operators to teach robots a set of physical movements has become commonplace and relatively cheap compared to separate programming.

The great advantage with automation technology is the flexibility of programmable workstations. Changeover costs from one batch to another are drastically reduced in terms of staff hours, special fixtures and tooling, and the downtime of production lines. Workstations are now programmed to incorporate sensor-actuated subroutines which automatically adjust to feature variations. Bar codes are used as a low cost actuation for other work characteristics. The use of sensors gives the flexibility to permit substantial reductions in inventories and work-in-progress.

9.7.3 Aggregate production effects

With automated plant, conventional personnel stoppages can be avoided and the benefit of higher equipment utilization taken. This can be transcribed into annual financial equivalents, but care is required in the forecast of the work load for such installations. An assumption of capability is no guarantee of realization.

It is at this stage that other incremental effects need to be considered and quantified wherever possible. A comprehensive check list is required of which the following are the more important items:

1. changes in quality costs, such as less scrap and rework costs. Where automatic inspection is part of the installation, quality control can be more rigorous at a lower labour cost;
2. equipment support costs, such as contract maintenance, the employment of new specialists etc.

9.7.4 Application development costs

Unless all the know-how comes from a specialist and proven supplier, the introduction of a new technology within its manufacturing operations will involve a company in application development costs. Typically, these include the planning, installation, commissioning and the reliability confirmation of the new plant, design costs for tooling and fixtures as well as software development expenses. Often this also includes the use of consultants. The firm will be familiar with some of these tasks but may know little about the embodied new technology. Even 'conventional' work has its hazards and surprises when the setting changes.

9.7.5 Proposal data preparation

There is the need to base plant performance estimates as much as possible on the actual conditions within the company, not on some supplier prepared material. Of course, such material could be helpful if based on systematic economic studies undertaken by the supplier, but they are no substitute for the context in which the company has to make its decisions.

The cost of the equipment is only a part of the total project investment. To this has to be added the cost of site preparation, installation and utility connections. More likely to be underestimated are the costs of systems integration, controlling, documentation and training accessories, installation and programming. For instance, a basic robot may only cost between 30–50% of the total investment involved with making it productive. The remainder covers such items as tooling, interfacing software and guarding requirements.

9.7.6 The evaluation of the transition period

When a company changes from a traditional manufacturing system to advanced technology, production has to be maintained during the transition period. A step-by-step approach over several years is often needed, even if everything goes well. Where problems are encountered in the application of the new technologies and systems the expected automation benefits are partial or delayed. The investment evaluation has to allow for this and to encompass both the running out of the existing plant and the running in of the new system. For a short period there can also be a duplicated manufacturing facility to ensure that customer commitments can be met and cash flow sustained.

9.7.7 The capacity implications of automation

Where automated manufacturing systems result in shorter production lead times, improved quality and other facets of customer value, extra sales are often expected by the company making the investment. There can be the temptation to anticipate additional contributions from this source. It is therefore important that such anticipation is supported by market research. Company sales staff are party to the economic success of automation.

9.7.8 Undue emphasis on direct cost savings

Many automation schemes have been put forward on the basis of savings in direct costs, particularly direct labour. These become the focus not only for evaluation but subsequent implementation. Other benefits to overall company performance are not always given the same level of attention and the focus on this particular aspect emphasizes the threat to jobs. This is not confined to shop floor workers; it also occurs with computer-aided design (CAD) where the prime justification is often seen as the reduction of drawing office staff, while the benefits of quicker technical estimates and reduced product development time are given less attention.

9.7.9 Change in the nature of labour costs

Whereas in traditional engineering production there was a direct link between machine and labour hours, the elimination of routine tasks and reduction in the overall labour content has resulted in a considerable reduction of direct labour costs. Furthermore, operator attendance on computer-controlled equipment is essentially of a multi-machine nature and often includes simple maintenance tasks, tool setting and systems surveillance. For a given equipment configuration shop floor labour has now the characteristics of fixed costs and is often treated as such.

9.8 INVESTMENT EVALUATION

In broad terms, investment evaluation is concerned with the economic justification of a proposition within a business context. This means, for instance, that assumptions made in the manufacturing context, such as the level of production, are confirmed by market prospects. Furthermore, the scope and nature of a proposition needs to be assessed. It is not unknown for a capital expenditure request to contain a number of related projects. The strength of those relationships need to be probed. Could individual projects be free standing and if so, what are their individual worth? Must they be carried out simultaneously?

The essence of the evaluation is the comparison between the capital expenditures to be undertaken and the cash flow benefits which consequently accrue. The justification is concerned with the recovery of the investment and the adequacy of the yield on the capital outlay.

The techniques of investment analysis contribute to the spectrum of management judgement. The very task of analysis requires a coherent and logical approach; the sifting of assumptions and a concentration on the salient features of a proposition. It should be an audit of the thought given to the proposition.. The scope offered by software packages makes computation a matter of seconds and facilitates rapid recalculations if different assumptions and scenarios are included in the proposition. Here is the opportunity for linking calculation with strategic thinking.

The actual techniques of plant investment appraisal have been well established for many years. On the whole, embodied advances in technology have been contained within an established framework of evaluation. This has been particularly so where technological progress was evolutionary rather than revolutionary; where the risks of new technologies contained in the proposed equipment and systems still seemed manageable.

9.8.1 The distinction between capital and revenue expenses

The categories of expenditure to be incurred are an important aspect of evaluation. The distinction between research costs, to be written off against the profit and loss account, and product development costs which can be capitalized, will be discussed in Chapter 10. There are the important aspects of tax liability and depreciation allowances to be considered as they can have a significant effect on the resultant project cash flow. It means that the annual project benefits have to be calculated net of tax.

9.8.2 The main techniques of investment appraisal

The major characteristics of the three main techniques are summarized below. Their mathematical computations are given in Appendix A. Each type has its family of variants. As they are essentially 'technology neutral' they are not developed here in detail. A fuller account of these methods is given by Lowe (1979). It should be remembered that the mechanics of computation give investment appraisal an aura of precision which the nature of the input data does not always justify. It is an important aid for decision making but remains at that level.

(a) The payback method

The concept of this method is simple: it measures the time from the initial investment until the net cash flow of generated benefit equals that outlay. The higher the cash inflow, the shorter the payback period, the more financially attractive is the investment in new technology. Its simplicity makes it attractive to many companies. It is easy to calculate and the payback time gives an apparent limit to the risks of the project. A management is always glad to know when its investment has been recovered. However, from an analytical point of view it has serious limitations. No allowance is made for the interest costs of the money outstanding nor for the project period after payback.

(b) The return on investment analysis

The horizon for evaluation is now the expected product or equipment life, not the payback period. The return on investment measures in percentage terms the benefits per annum, after depreciation, against the original investment outlay. The provision for depreciation has to incorporate the life expectation of the product or equipment. This is affected by future market and technological changes. If obsolescence comes before the assumed lifespan it will be necessary to write off the outstanding depreciation. In contrast to such an unpalatable decision, if the accumulated depreciation correspondingly reduces the book value of the investment, say, in equipment, a constant stream of benefits will yield an increasingly attractive return. The temptation then is to forego a strategic view and keep aging plant for longer than the company's total situation warrants.

(c) The discounted cash flow method (DCF)

This is the most appropriate technique for any substantial project with an extended time-scale. It is applicable to R&D projects with a time series of expenditure before there is any yield from a project. Discounting on a time basis is a key feature of DCF. It allows, in the form of compound interest, for the time lag between actual disbursements and subsequent receipts. In this manner there can be a valid comparison between capital expenditure now and future revenue benefits. With long time series of cash flows the more distant future is more heavily discounted. While it becomes progressively more difficult to forecast the cash flows as the time span is extended, the technique also compensates for lower accuracy by discounting estimating errors.

9.9 PROJECT RISK ANALYSIS

Risk analysis applied to R&D projects has already been mentioned in Chapter 8. Here, risk is seen in a total context of project outcome, i.e. all other possible risks are envisaged. In essence, risk relates to project outcomes which are different to that predicted. Seldom does a project perform financially to the precise figures contained in a single outcome calculation. Essentially, such a computation reflects the most likely outcome as envisaged by the project proposers. Management may prefer to know the degree of this likelihood and the range of other possible outcomes. Better than expected results will always be welcome, but it is the risk of shortfall which makes for attention.

As with R&D, some of the risks with major equipment investments depend on the degree of novelty these incorporate. Novelty here also includes the nature of the application by the prospective user. Where the application of the new technology is critical to the competitiveness of the company the risks appertain not only to the investment itself but also the company's market position.

Where a risk analysis is a particular feature of the evaluation process, a number of scenarios are often explored, each with its most likely outcome and associated risks. The evaluation context can be structured in the following typical manner:

1. the choices available; e.g. which process, what plant size?
2. the factors which influence choice; e.g. plant reliability;
3. the range of possible values for each factor;
4. the effect of factor value changes;
5. the probability of each factor value.

The behaviour of certain key factors requires special attention. Consider, for instance:

- project delays,
- patterns of expected sales,
- reduced capacity working,
- plant maintenance and downtime problems.

The techniques of risk analysis are well established and can be applied to suit the context of a particular investment proposal. Only a summary is given here but the interested reader can find a fuller account by Lowe (1979). These techniques are supplementary to those reviewed in Chapter 8.

9.9.1 Assumed certainty

Here the hazards of the the 'most likely' estimates are recognized. Every key value and assumption is scrutinized and adjusted until the analyst is certain that each project parameter cannot go beyond the stated value. For instance, the most likely man/hours for a design/development project will be 100 000, but the responsible manager will be quite certain that 135 000 will be the limit. The overall effect of this form of scrutiny will be that management is satisfied that the project outcome will not be worse. The approach is similar to the adding of safety factors in engineering design, but the compounding of caution may be such that the project is not accepted.

9.9.2 Risk adjusted rates of return

Where a company has a basic rate of return, which is a minimum requirement for any significant expenditure project, a further percentage requirement is added as a form of cover against the risks of a proposition.For instance, if the company base rate were 15% and because of the perceived project development risks a further 5% was required the project benefits would have to yield a return of 20% on the investment.

9.9.3 Multi-level estimates

A common example is the three-level evaluation, based on the most optimistic, pessimistic and most likely combination of project factors values. The extremities of expectations give a spread of possible project outcomes.

The weighted three-level estimate is a development of the previous method. By definition the 'most likely' outcome has a greater chance of being realized and it can therefore be argued that it should be give greater weight. A mean outcome value is computed on the basis that the most likely outcome has four times the weighting of the other forecasts.

9.9.4 Sensitivity analysis

Another way of examining possible variations in a project outcome is the application of sensitivity analysis to some of the factors which determine the project yield. Consider the following typical factors with a new technology process:

- project development cost,
- level of capacity working,
- labour costs,
- product price, etc.

After the determination of the single value DCF return rate, based on the 'most likely' factor values, attention is then concentrated on the range variation of a particular factor, with all other values held constant. If the effect on the calculated rate of return is substantial then the project outcome is sensitive to the variations of that factor. The project outcome will generally be very sensitive to one or two factors and insensitive to most others. This is, of course, important for project estimating and subsequent project implementation. Management attention needs to focus on those factors to which the project is hostage. Figure 9.3 gives a telling example how one automation project proved to be vulnerable to the level of capacity working in an engineering factory.

9.9.5 Stochastic evaluation

The patterns of evaluation described so far have been mainly deterministic. A further refinement of the evaluation process is to take a number of likely project scenarios and to attach different probabilities to them. This is the stochastic approach. For instance, we can take different product development time scales or levels of subsequent product market shares. For each combination of assumptions, a probability value is given, a cash flow profile is established and the project outcome calculated. From the product of these outcomes and their probabilities a mean project value can be computed. An important statistical measure of risk is the spread of project outcomes which is given by the coefficient of variation for the portfolio of alternative outcomes. An example of this form of evaluation is given in Appendix B.

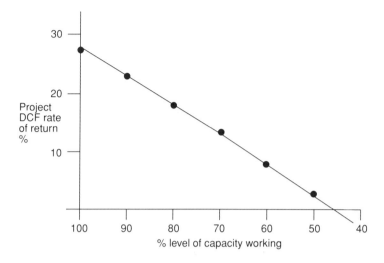

Figure 9.3 Project sensitivity to the level of capacity working.

9.9.6 Simulation

With the previous method a scenario consisting of a pattern of assumptions was given a probability value. Now each determinant of the project outcome is given a probability value. This allows specialist judgement to be integrated more effectively in the overall outcome. For instance, Table 9.2 shows the judgement of a technical director about the likely development costs for a new product.

Table 9.2 An estimated range of likely product development costs

Expected development cost range (£1000)	Range probability	Cumulative probability
650–750	0.01	0.01
750–850	0.05	0.06
850–950	0.10	0.16
950–1050	0.40	0.56
1050–1150	0.25	0.81
1150–1250	0.15	0.96
1250–1350	0.04	1.00

Similar probability assessments can be made about the other project outcome determinants such as the costs of production equipment, market volume, likely product life etc, each assessment reflecting the judgement of the staff most experienced in that function. The result will be a number of tables with a range of values and their corresponding probabilities.

With the application of the Monte Carlo technique, using random numbers, a particular pattern of project values will be generated. From this a
project outcome will be computed. As with this technique the frequency
of a particular value coming up reflects its probability, due regard will be
paid to each value judgement. With a PC and a simple software package
a large number of iterations can be undertaken and the results can be
shown graphically in a project profitability profile. Figure 9.4 illustrates
this. It not only gives the modal (most likely) outcome values but shows
the implications of the tails of the distribution. What matters to management is not only the likely outcome but also the degree of risk of an
unfavourable combination of project factors.

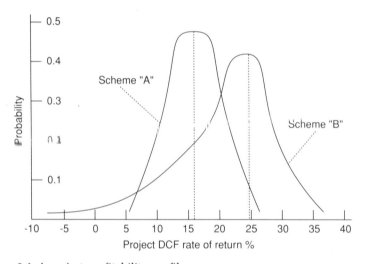

Figure 9.4 A project profitability profile.

9.9.7 Decision trees

Where a corporate technology strategy requires the implementation of a
series of technology investments and subsequent investments are determined by previous projects, the various scenarios can then be explored
and related to each other by the development of a decision tree, such as
shown in Fig. 9.5. In this example there are two decision stages. The first
stage is to develop the process and the second to build a plant using it.

The two main imponderables are the outcome of the development project and the behaviour of competitors. Probability values can be assigned
to each scenario and the worth of decision alternatives calculated. Likely
competitor behaviour can be mirrored by different scenario probabilities.
For instance, it is quite possible for competitors on the same opportunity

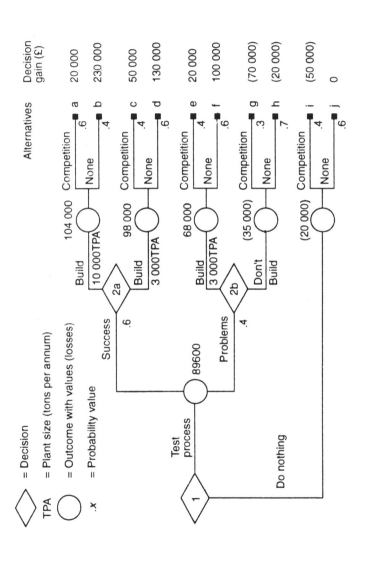

	Alternatives	Decision gain (£)
◇	= Decision	
TPA	= Plant size (tons per annum)	
○	= Outcome with values (losses)	
.x	= Probability value	

		Decision gain (£)
Competition .6	a	20 000
None .4	b	230 000
Competition .4	c	50 000
None .6	d	130 000
Competition .4	e	20 000
None .6	f	100 000
Competition .3	g	(70 000)
None .7	h	(20 000)
Competition .4	i	(50 000)
None .6	j	0

104 000 — Build 10 000TPA

98 000 — Build 3 000TPA

2a — Success .6

68 000 — Build 3 000TPA

(35 000) — Don't Build

2b — Problems .4

(20 000) — Do nothing

89600 — Test process

1

Figure 9.5 Decision tree of investment in new technology.

track to experience similar development problems and become cautious about market entry. It will be noted that to do nothing is also a decision, even if it is arrived at by default. Alternatives 'g' and 'i' show that the advantage then is with the competition.

The value of the decision tree method is that it forces a management to look at all scenarios systematically. It brings into the open what otherwise might not have been considered.

9.10 SOME PROBLEMS OF INVESTMENT EVALUATION

The mechanics of evaluation tend to obscure some of the difficulties of judgement. Management has to be aware of the following aspects.

9.10.1 The presumed fixed position

One of the most dangerous assumptions which underlies many investment appraisals is that the base line of appraisal remains constant. The decision maker looks at the proposition as if it were from a fixed point. He looks mainly at the incremental aspects of the proposition and assumes that if the investment is not authorized the *status quo* will persist. Too busy is the company where this is so.

Relativity is not just advanced physics: it also applies to business. What happens when a firm decides not to invest in new technology? (If it does not even think about it then its days are numbered.) It will gather 'operating inferiority' and experience a gradual erosion of its competitive advantage which, in turn, will result in loss of market share and cashflow. Where there is rapid technological change it will quickly become uncompetitive and its order book will dry up. The company products have slipped down the obsolescence cliff. The implications are shown in Fig. 9.6.

9.10.2 The shortcomings of accounting systems

The well-known forms of investment appraisal: discounted cash flow (DCF), payback and return on investment methods all rely on existing accounting systems and conventions for their information. As will be mentioned in Chapter 10, these systems usually emphasize savings in direct costs, particularly labour costs, but become less explicit about some of the repercussions on indirect costs, where problems of measurement often lead to an 'other things equal' approach. The most significant, yet hidden, assumption is that only the incremental effects of a proposal need to be considered. The term 'incremental' is usually taken as accountancy reportable. For the strict purpose of analysis everything else is presumed to stay as it is. Unfortunately, in many competitive industries this is the least likely of future scenarios.

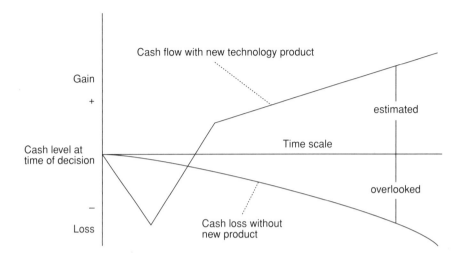

Figure 9.6 The cash flow implications of the investment decision.

9.10.3 Incorporating intangibles in the investment evaluation

It is important to segregate claimed, proposition-related benefits, such as reduced manufacturing lead time or higher quality, from a specific technology investment proposal. Of course, they are valid aspects in a decision context. But should these intangibles not be achieved anyway, irrespective of the particular project?

Take the matter of reduced delivery lead time. Below a certain competitive value which a company may be expected to achieve anyway, even without an investment in new technology, what value is added by a further reduction? Another aspect is illustrated by the following example. A case was made by the manufacturing engineers of an engineering company that a new numerically controlled machine tool would simplify the introduction of component modifications. When pressed to put a value on the stated benefit, none could be established. The reason? No previous modifications could ever be recalled.

It is an accepted practice that companies value certain intangibles, such as product brand names, which can appear as assets in their balance sheets. This practice is not without debate, but a brand name is accepted as an asset of economic worth in terms of the discounted value of future cash flows that can be attributed to it. This approach to intangibles which accountants accept on one side of the business could usefully be applied to the intangibles of an investment proposal. That it is not done reflects organizational behaviour, not principle.

One way of incorporating the effects of intangibles is to prepare *pro-forma* profit and loss accounts and balance sheets for several years forward. These can illustrate possible scenarios of performance based on the likely values of such key factors.

9.10.4 Market-based intangibles

The effects of such intangibles, as improved delivery or better quality, provide market opportunities the extent of which can be assessed in terms of extra sales, or during a recession, the retention of a given market share. It is not an easy task but it needs to be done. While no high accuracy can be expected a set of three-level estimates can be made and integrated in a range of outcome scenarios. The marketing department is a participant in the investment preparation process. Its experience and expertise are relevant constituents of company judgement. If necessary, appropriate market research has to be undertaken. One approach is a form of structured citation analysis of customer views, particularly those linked with business that went to competitors. To do nothing or to guess does no justice to the rigour that can be expected in the evaluation of the investment process.

In the context of quality and reliability, a UK based car manufacturer, recently commissioned a customer satisfaction survey. One of the questions it asked customers was whether they would, in due course, buy another car from the company. The 'yes' response was 44%. The same question, when put to the customers of one of its chief competitors, gave a 70% 'yes' return. Although the survey results were no more than indicative, the company considered the findings to have a 10% tolerance. This could be transcribed into a likely loss of turnover, which in turn could be expressed in loss of cash flow.

9.10.5 Trade-offs between intangibles

As part of the evaluation of intangibles it is possible to establish their relative importance. This can be done on the basis of a pairwise comparison scale. If two intangible benefits, such as shorter lead times and improved quality are compared, their comparative importance can be ranked. For instance, on a scale of 1 to 5, the value 1 means that they are regarded of equal importance, whereas 5 indicates that quality is of absolute importance compared to lead time. The various intangibles can then be structured within a matrix and their overall values calculated and judged. If the row in the matrix is improved quality its value on the shorter lead time column would be 5. Correspondingly for the lead time row the improved quality column would be 1/5. On that basis dissimilar sources of benefit can be compared in relative importance. While it is appreciated that this technique does not provide actual financial values, it does spell out the relative importance of particular benefits. This justifies more detailed investigation and control of the source of benefit specified.

9.10.6 The evaluation of automation projects

A major automation project, like any other substantial project, must be seen within the context of company strategy. The longer term view, which this expresses, needs to stress the benefits of the proposal rather than the technology to be used. The benefits must be reckoned to accrue to the company as a whole, not only to a department.

The first point to appreciate with such an investment is the commitment to a hardware/software system. The longer-term aspects of systems compatibility and standardization require policy decisions which could go well beyond the proposal to be evaluated. These considerations become more acute where a company is also considering the installation of a computer-aided design (CAD) facility or a computer-integrated manufacturing system (CIM).

9.10.7 An 'option' approach to advanced technology

Busby (1992) has an interesting approach to the investment in advanced manufacturing technology. He sees the investment as a source of cash flow which also yields further technological opportunities that can generate subsequent cash flows. Although not necessarily well defined at the initial stage, these subsequent opportunities have value and Busby regards them as 'options' as they allow a firm to take up such prospective developments. Like options in the financial markets, they have to be bought. With manufacturing technology, for example, this could typically take the form of investment in computer-controlled systems. The value of the option is the flexibility it offers but the company has to decide what price to pay for it.

9.10.8 The dilemma of inflation

For a company, inflation is both unpredictable and uncontrollable. The incorporation of inflation in the decision framework for the investment in new technology is a contentious matter. A number of possible approaches have been summarized by Lowe (1979). To ignore inflation in the management of a business is a prescription for bleeding to death. Yet to make a guess as to what inflation might be several years hence is to accept a hazard blindfolded. What is worse in such a case is the erosion of analytical discipline. If you make such a guess, why not guess the behaviour of other factors? There is also the particular effect of rapid technological change. For instance, computers and computer systems have become cheaper in real terms and within a composite plant investment proposition can mask different equipment cost trends.

One practical approach is to take the investment proposition at current costs and then adjust the investment parameters annually at the rates of inflation actually experienced. This could be part of standard operations within a framework of current cost accounting. Equipment would be revalued in real terms and depreciation would be charged accordingly, just like other inflation adjusted operating costs. Such a method would then yield an inflation-adjusted cash flow which helps to maintain the real worth of the company. It is appreciated, of course, that to maintain corresponding product prices invites competitors to take a greater market share, but that consideration is part of a marketing strategy and should not confuse investment analysis.

9.11 NEW TECHNOLOGY AND THE PLANT REPLACEMENT DECISION

An important, although less emphasized avenue of technology advance is the routine plant replacement decision. There is a considerable amount of literature on plant replacement and a number of models have been evolved. Terborgh (1949, 1950) and the Machinery and Allied Products Institute (MAPI) provide a useful framework for assessing the effects of future technological change. Their concept of the 'operating inferiority' of existing plant consisted of the summed effects of deterioration and obsolescence. The effects of deterioration of manufacturing plant were relatively easy to establish by adding the cost of maintenance and downtime losses. Obsolescence losses reflect the benefits of embodied technological advance, not adopted by the firm but by its competitors. It is essentially the loss of competitive advantage. Such costs are difficult to measure, as they are not readily reported by any accounting system. Terborgh estimated these by an annual 'inferiority gradient', an approximation that was defensible as long as technological change was modest. The concept of operating inferiority is shown in Fig. 9.7. When there is a major technological change, such as with automation, the obsolescence gradient becomes a cliff. Efficient competitors who have adopted automation schemes will gradually undermine the company's market position with products at lower prices and a better response to customer requirements. Other things being equal, the company's order book will shrink and all the diseconomies of scale due to a lower turnover will affect its costs of production.

Of course, the company cannot buy new equipment every year just to catch the latest technical improvements. In most cases, the capital cost would be much too large, never mind the continuous disturbance to operations. A balance therefore has to be established between the capital

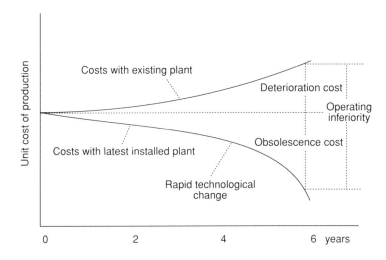

Figure 9.7 The concept of operating inferiority.

outlays and the operating cost stream. The balance is achieved by selecting suitable intervals of plant replacement. The optimum interval is that which minimizes the total discounted capital and operating inferiority. With a slow change of technology, obsolescence will only slightly increase the relative production costs and the replacement interval will be correspondingly longer.

9.12 THE AUDIT OF NEW TECHNOLOGY PROJECTS

The project outcome consists of all the incremental changes that can be ascribed to the advent and completion of a project. The assessment of the outcome can be in two ways:

1. the evaluation of quantitative and qualitative changes,
2. a comparison between project promise and project outcome.

The audits of such projects have a number of challenges and much depends on their boundaries. There is often a spill-over with development work into related activities, which can distort the costs and benefits of a particular project. It is relatively easy for a company to establish whether a project has been completed to schedule and within budget. One of the great problems with development projects is, of course, the propensity for cost overruns. As we have already noted previously, the greater the technology jump, the greater is the risk of technological problems and cost overruns. One key audit ratio is actual project cost/estimated project cost. With high technology projects, particularly in defence or aerospace, such ratios have been as high as 4:1. With well established technologies and little modification the outcome may be 1:1. Where the project duration is more than, say, one year, adjustments need to be made to the cost ratio to allow for the effects of inflation.

The comparison between final and estimated costs is often complicated by changes in project objectives that occur during the development period. It is then difficult to isolate the corresponding adjustment costs from a multitude of other cost-influencing factors.

The market performance of a new technology product will be a relentless audit of the project. In essence, the customer is the auditor. With new technology plant, the claimed benefits can be related to actual performance, such as output rates, manning levels, process yields and downtime. The timing of such audit is important. If it comes immediately after product launch or plant commissioning there can be the distortion of starting problems. If left too long then subsequent developments tend to obscure the process outcome.

9.13 SUMMARY

Investment in new technology is in product and/or process technology. Its strategic assessment requires both an investment evaluation and a scenario cost/benefit analysis. Estimating for development projects is based on the definition and content of the work tasks involved. Staff costs are often the major part of the development costs and form the focus for financial risk reduction.

Investment in new technology processes requires a hierarchy of evaluations, from the detailed production task to the overall business impact, before the expected economic benefits can be ascertained. The three main techniques of investment appraisal are the payback method, the return on investment analysis and the discounted cash flow method. A range of techniques are available for project risk analysis. Decision trees can be used for more complex, multi-stage investment programmes.

Management needs to appreciate some of the underlying problems of analysis, such as the nature and fragility of some data, the valuation of intangibles and the dilemma of inflation. Above all, it must resist the assumption of the 'fixed position'; that the *status quo* will continue if it decides against an investment in new technology.

FURTHER READING

Bussey, L.E. (1978) *The Economic Analysis of Industrial Projects*, Prentice-Hall, Englewood Cliffs, NJ.

Hackney, John, W. (1992) *Control and Management of Capital Projects*, McGraw-Hill Inc., New York.

Lowe, P.H. (1979) *Investment for Production*, Associated Business & Press, London.

Massé, P. (1962) *Optimal Investment Decisions*, Prentice-Hall, Englewood Cliffs,NJ.

Primrose, P. (1991) *Investment in Manufacturing Technology*, Chapman & Hall, London.

Rose, L.M. (1976) Engineering Investment Decisions, *Elsevier Scientific Publishing Co.*, Amsterdam.

CHAPTER 10

Accounting for technology

10.1 INTRODUCTION

Accounting and finance permeate the management of technology. We have already noted in Chapter 3 that engineering economics describes the decisions made by technologists on the basis of cost data. Chapter 6 referred to the provision and use of R&D funds, while Chapters 8 and 9, were very much concerned with the assessment of costs and cash flows.

The purpose of this chapter is to consider further the accounting and financial aspects of technology and technological change. This includes both financial and management accounting and involves:

- the accounting for R&D expenditures,
- the treatment of technology as a balance sheet asset,
- the role of accountancy in technological competition.

Accountancy is a vast subject and within the scope of this book only those dimensions relevant to the theme of technology can be described.

It is worth mentioning here that long before the advent of computer-based automation the first successful commercial computer was installed in 1947. This was in the accounts department of a large UK catering and retail shop enterprise, where the operations of a physically massive unit were applied to the sales ledger. One of the first triumphs of advanced technology and computers was its successful application to accounting operations. Technology is not new therefore to accounting, yet the adjustment and use of accounting techniques to assist a company in securing and maintaining a competitive advantage is relatively recent.

10.2 ACCOUNTING FOR RESEARCH AND DEVELOPMENT

The acquisition of new technology, which is one major purpose of R&D, is a material and essential part of the activities of many substantial companies. The scale and the cost of these activities can be appreciated when in some industries, such as electronics and aerospace, R&D expenditures can be as much as 15% of company revenue.

Apart from Batty (1988) who endeavours to integrate cost accounting principles with the management of R&D, the topic is not well covered in accountancy texts which concentrate much more on the other operational functions of the firm. This is significant, bearing in mind the risks of heavy cost overruns with R&D projects. Although there is no precise boundary between R&D and DDMM, the costs of the latter are generally budgeted within, and absorbed as part of, the overheads of the business. Where these are essentially pre-production costs they are treated as operating expenses and charged against the new product.

From a financial accountancy point of view, the treatment of R&D expenditures can have significant repercussions on the company's accounts. It therefore becomes a concern of the board of the company. The professional view is encapsulated in the Statement of Accounting Practice (SSAP) 13 *Accounting for Research and Development* published in 1989 by the Institute of Chartered Accountants in England and Wales. It draws a distinction between research expenditure, aimed at gaining new, or advancing existing, scientific and technical knowledge, and development expenditure, aimed at using such knowledge for specific commercial projects. This, for example, can include the development of new materials, devices, products or services, as well as the installation of new processes or systems.

Apart from the expenditure on fixed assets, all other research expenditure should be written off against the profit and loss account as it is incurred. For development expenditure a firm has the choice between a write-off policy and a deferral policy. Where the latter is chosen the development expenditure will be carried forward as an intangible fixed asset in the balance sheet and will be written-off against future related revenue. The argument for such a capitalization is based on the yields expected from the developed product. A company needs to consider the implications of its accounting policy. Write-offs can make big holes in the profit and loss account; deferrals tend to 'smooth' these effects on the reported profits which may be an important aspect particularly for institutional shareholders.

The consequences of the choice can be profound, because of its effects on the balance sheet and profit and loss account of the business. In some cases the financial standing of the company becomes hostage to it as it affects the attitudes of major institutional shareholders and banks. This can have repercussions on the financing of R&D expenditure.

The critical nature of this subject has been well documented in the investigation report into Rolls-Royce Ltd under Section 165 (a) (i) of the Companies Act 1948 by the UK Department of Trade and Industry. In essence, the company's development of its RB 211–22B aircraft engine proved much more costly than anticipated. This expense was capitalized; had it been charged against the profit and loss account, the company would have had to report heavy losses. That might have given an early, albeit uncomfortable, signal of the disaster to come. As it was, it drained the company's financial resources until it was forced into receivership. Its subsequent rescue by the UK government cost many millions of public money. Eventually, however, the RB 211–22 engine family became a great success as a world class product.

There are several requirements which have to be met by a company before the capitalisation of development expenditure. The following are the most important:

1. There must be a clearly defined project.
2. Related expenditure must be separately identifiable.
3. There must be reasonable certainty about the technical feasibility and commercial viability of the project.
4. Future revenues from the project must be 'reasonably expected' to exceed the deferred development costs and other future related costs.
5. Adequate resources must be 'reasonably expected' to be available to complete the project including such working capital as may be required.
6. The deferred capital expenditure must be amortised over the expected period of commercial exploitation. If future expenditure recovery becomes doubtful then appropriate write downs must be made.

10.2..1 The disclosure of research and development expenditure

By virtue of the UK Companies Act 1985, Schedule 7, Part II, the directors' report shall contain an indication of the activities (if any) of the company and its subsidiaries in the field of research and development. This is a very general requirement but it is reinforced by Schedule 4 of the Act, 'Form and Content of Company Accounts'. Part 1 of the schedule deals with the balance sheet formats. Under the heading of fixed assets, the following intangible assets have to be stated:

• development costs,
• concessions, patents, licences, trademarks etc.

Correspondingly, the formats for the profit and loss account require a statement of the depreciation and other amounts written off tangible and intangible fixed assets. It is interesting to note that only development, and not research expenditure can be accounted for in this way.

Expanding on these basic legal requirements, SSAP 13 requires that the accounting policy relating to research and development be stated and explained in the accounts of the company. Deferred development expenditure should be shown in the balance sheet as an intangible fixed asset. Movements of such expenditure and the amounts carried forward at the beginning and the end of the period must be disclosed.

10.3 TECHNOLOGY AS A BALANCE SHEET ASSET

There are the cases where a technology is acquired, typically in the form of patents or licences, rather than developed within the company. The expenditure is then regarded as an investment which will also be shown as an intangible fixed asset on a company balance sheet. Its value is primarily determined by the expected income stream such an asset generates. While the acquisition cost is on record, the subsequent values of the asset become a matter of judgement. In any case, the asset will have to be written off by the time the patents or licences expire.

Where there is doubt about the continued value of the asset its write-off will have to be accelerated. This could happen when technological change erodes the expected revenues. It is therefore necessary to monitor this asset, its use, economic life and the life cycle position of the technology. Special provisions will have to be made where key patents have been infringed or are subject to litigation.

10.4 THE ROLE OF MANAGEMENT ACCOUNTING

When a company aims to achieve a competitive advantage, either by added product value or by cost leadership, it needs a data framework to set targets and to assess performance. For this an effective management accounting system has to be in place.

It has already been noted in Chapter 3, that much of technology decision making, such as the choice of processes and their related equipment, depends in part on cost data. The quantified benefits and costs are marshalled and incorporated in a judgement. This is the context for process planning, plant investment or replacement, and many other decisions for which industrial management is responsible. As cost data are an essential ingredient for these it is important to assess their value and suitability for such purposes.

The main sources of such data are the financial and management accounts of the company. With most industrial firms the financial accounts express commercial needs and the statutory requirements of company law. They have not been designed, as such, to assist management in making decisions about new technology and the equipment or systems in which such technology is embodied. They are not well fashioned for the valuation of innovation, ideas, risks and intangibles which, in a competitive industry, may be just as important as plant and buildings. Similarly, management accounts and the related techniques of estimating reflect the context of existing manufacturing activities.

For instance, standard costing relates to mass production within a relatively stable production environment and significant direct labour costs. It was feasible in such a situation to operate an absorption costing system with direct and indirect, or fixed and variable production costs. Typically, overhead charges were expressed in percentage terms of prime or labour costs and focused attention on the reduction of the latter. When a process was labour intensive and there was a large direct work force, labour efficiency improvement could reap substantial returns. The introduction of automation and flexible manufacturing has made small batch production feasible and in many cases has drastically reduced the labour content of production. The absolute gain of improvement in labour efficiency has correspondingly declined. Furthermore, to evaluate such new systems by techniques which carry the assumptions of a different manufacturing context increases the risk of misjudgement. In many industries the direct labour costs are now less than 10% of the total manufacturing costs and are an even smaller proportion when a company's non-manufacturing costs are included. To build a costing structure on such a modest base can introduce substantial bias.

With many high technology manufacturing companies overheads have become so high that they obscure true product costs.

10.5 THE MAIN COSTING SYSTEMS

It will be appreciated that the nature of a company's costing system depends on its operations. In turn, a costing system influences judgement and the way decisions are made. According to the engineering manager of a well-known US electrical company: 'the way in which we allocate overhead costs affects the way in which we design products'. It is therefore appropriate to briefly mention the most important systems in use and to assess their relevance to the challenges of technological change. The first two systems are primarily conventional cost reporting systems. The other three are more recent and reflect the pressures of world-wide industrial competition; they are tools of cost reduction. These five systems are briefly reviewed below but readers who wish to go into more detail are referred to the many textbooks of cost accountancy some of which are listed at the end of the chapter.

10.5.1 Absorption costing

The prime function of this system, as the name implies, is a framework for the recovery of overheads. It is also used as a basis for pricing and estimating. The apportionment of overheads are usually based on prime costs or one of its constituents, such as direct labour or material costs. It is essentially an inward-looking, mechanistic technique which may pay insufficient attention to the opportunities and challenges of the market place. It does not track the sources of competitiveness, such as quality and flexibility. Figure 10.1 illustrates the absorption costing method.

Consider a firm manufacturing instruments. For the next financial year it is budgeting for a turnover of about £7M based on the sale of 10 000 instruments. The following is a simplified budget statement based on absorption cost accounting.

Cost budget
Prime cost

	£ M
Direct labour	0.8
Bought-in-parts and materials	0.9
Direct expenses	0.1
Total prime cost	1.8
Total indirect costs	4.0
Total budget cost	5.8

Pricing structure for 'average' instrument	£
Direct labour	80
Bought-in-parts and materials	90
Direct expenses	10
Prime cost of instrument	180
Indirect costs, as 500% of direct labour costs	400
Total unit cost	580
Profit, added as 20% of total unit cost	116
Instrument price	£696
Total revenue budget	£M 6.96
Overall profit budget	£M 1.16

Figure 10.1 Absorption costing – the mechanistic routine.

Consider the same firm using a marginal costing approach.

Cost budget

	£M
Total variable cost	1.80
Total indirect costs (overheads)	4.00
Total budget costs	5.80

Sales budget

Instruments	Price £	Revenue £M	Variable cost £M	Contribution £M
4000	710	2.84	0.72	2.12
4000	700	2.80	0.72	2.08
2000	680	1.36	0.36	1.00
10 000		7.00	1.80	5.20

Overall profit budget

Overall profit = overall contribution − total indirect costs
= 5.20 − 4.00 = £1.20M

Figure 10.2 Marginal costing – the commercial approach.

10.5.2 Marginal costing

With this technique, variable costs are subtracted from sales revenue to give a contribution which is used to cover overheads. If at the end of a trading period there is a surplus this constitutes a profit; if the contributions are insufficient to cover overheads then a loss will be incurred. This technique is often used by companies which have capital intensive processes. Marginal costing puts more emphasis on marketing judgement. There may be no detailed evaluation of indirect costs. Figure 10.2 shows the use of marginal costing.

10.5.3 Activity-based costing (ABC)

It is basic that a business will only stay competitive if it provides value to its customers and that its costs are less than the price they are prepared to pay. Activity costing, and for that matter activity management, will be concerned first with the sources of such value. They will also be concerned with the non-value adding activities, the cost drivers, which are traditionally subsumed in the overheads. These are typically: materials handling, change orders, etc; any activity which takes significant staff resources and yields little value. In the light of established activity costs, management can make effective operational decisions. The management information system focuses on the control of those activities that determine the competitiveness of the business.

In his investigation of a US General Electric assembly line, Johnson (1990) indicates how activity costing can be used to re-establish the competitive position of a manufacturing plant. The following tasks are involved:

- Identify externally focused measures of how the product gives value to the customer and how the product line supports the goals of the company.
- Identify activities that cause work on the product line and assess how each activity adds value or creates waste.
- Identify and eliminate generators of the work that cause non-value activities.

In his further study of an Allen-Bradley plant Johnson describes how the allocation of departmental indirect costs is carried out by departmental managers according to their estimates of how departmental activities will generate costs over the next six months. Simple PC spreadsheets readily permit such calculations to be made. While the plant accounting system allocates labour hours and costs to the department it does not allocate overheads below departmental level. In substance, therefore, the judgement and decision about cost drivers is made at local level. This facilitates better local control and accountability for results. The emphasis is on the control of the cost drivers. Shop floor staff have a better understanding how their activities affect costs and how they can appraise their own competitiveness.

The main features of such a system is a set of cost drivers which are established after the careful cost analysis of existing operations and then formulated to be meaningful to product designers. For instance, Cooper and Turney (1990), reporting on a costing system introduced at a Hewlett-Packard plant for the assembly of electronic equipment, noted that a machine-hour was insufficient for such a purpose and listed the following cost drivers with appropriate data:

1. number of axial insertions,
2. number of radial insertions,
3. number of dip insertions,
4. number of manual insertions,
5. number of test hours,
6. number of solder joints,
7. number of circuit boards,
8. number of parts,
9. number of slots.

It was not the primary purpose of the costing system for designers to achieve an absolute minimum cost design at the expense of other product attributes, but designers had to have a feel for the cost implications of their designs. For instance, a manual insertion cost rate was 2–3 times the rate of an automatic component insertion.

One of the strongest reasons for activity based costing has been the need to stay competitive. More accurate product costs yielded better pricing and product mix decisions. The cost system was designed to improve the external marketing strategy of the company.

With rapid technological change and shorter product lives there is greater pressure to reduce product costs through improved design and manufacturing operations. An activity-based system for such goals is designed to reduce product development lead times and to emphasize manufacturing cost rates at the design stage. Such systems focus on manufacturing capability.

10.5.4 Target costing

This is a cost management tool first introduced by Japanese companies to reduce overall product costs over the entire product life cycle. The product must make its contribution to the long-term profitability plan based on a target profitability index. Following detailed market research surveys, the sales plans and target product price are established. A target profit is subtracted to give the target cost to which the product has to be designed and manufactured. There is continued, unremitting emphasis on value engineering. This approach has a cost reduction driver and pays much more attention to development and design activities. Standard values are set for labour, material and parts costs as a basis for period cost improvement programmes. Figure 10.3 indicates the nature of target costing.

The same firm, with a new management, adopts a target costing system. Following a strategic benchmark evaluation, the sales manager states that 12 000 instruments can be sold if the average price can be reduced to £600. The company has a profit target of £1.5M for the budget year. The following is its outline budget statement.

	£ M Overall	£ per unit
Revenue		
12 000 instruments at mean price of £600	7.20	600
Target profit	1.50	125
Total allowable costs	5.70	475
Allowable direct labour	0.90	75
Allowable bought-in-parts and materials	1.05	87.5
Direct expenses	0.10	8.3
Total allowable prime costs	2.05	170.8
Total allowable indirect costs	3.65	304.2

Unit cost reduction targets (£)		
(see also Fig. 10.1)	From	To
Direct labour	80	75
Bought-in-parts and materials	90	87.5
Direct expenses	10	8.3
Indirect cost reduction target (£M)	4.0	3.65

Figure 10.3 Target costing – the strategic drive.

10.5.5 Life cycle costing

Although this technique has been in use for some years for equipment investment evaluation, it is becoming increasingly relevant to product costing. Life cycle costing is concerned with the totality of product costs during the product life, not just those of a particular stage. In costing terms the product life cycle starts when company resources are devoted to it. That may be a considerable time before the product is launched and makes a contribution to the company revenues. The life cycle ends when the product is withdrawn and all residual costs have been ascertained. As product lives are becoming shorter heavy development and launch expenditures have to be recovered more quickly. This includes the depreciation costs of all product-specific equipment and tooling.

10.6 PROBLEMS WITH COSTING SYSTEMS

Technological progress, particularly manufacturing automation, has resulted in the following:

1. An increase in the capital intensity of production and the reduction of unit labour costs.
2. The distinctions between direct and indirect production labour and their costs have become blurred. With some companies production labour has so diminished that it has become a residual, fixed cost.
3. With technological change the expected equipment life is often less than the write-down provisions allowed by taxation. If the depreciation charges are substantial this could lead to major distortions in cost allocations.
4. There is a greater need for life cycle costing with shorter product lives.

The investment in new technology and the management of improvement, which often involves technical changes of a minor nature, require cost data for assessment and decisions. Suitable cost reports are essential for cost reduction programmes as much of their impact would otherwise be lost. This has been of concern to a number of companies which have reviewed their management accounting processes and financial systems as part of overall change programmes. If task forces are appropriate for cost reduction projects than a similar approach is pertinent to ascertain the efficacy of the related measuring and reporting systems. Where a manager is accountable for the success of new technology then he is entitled to have a costing system that strengthens his endeavour.

10.6.1 The distortion effects of different costing systems

Where a conventional accounting system is used, Banker *et al.* (1990) have found that in plants which manufacture a wide range of products, the more complex products tend to be undercosted. They came to this conclusion by the comparison of original overhead costs assigned to a product and by comparing this with the traced costs established by an activity-based cost system. This took into account the time spent by indirect personnel on the products concerned. A major source of difference was the cost allocation for supervision staff based on the premise that the function of such staff was 'to supervise direct labour' and that the mechanistic allocation of overheads was adequate, irrespective of how such staff actually applied their time to various products. Considering that good supervision involves problem solving then the complexity of products was a significant variable.

10.6.2 The alignment of the costing system to company strategy

As we have seen, traditional costing systems, such as absorption costing, are essentially mechanistic in operation. There is an implied assumption of stability underlying a system and its operations. In a competitive context, where all the resources of the company need to be marshalled to achieve success, or even just to survive, it is important for a costing system to emphasize the role of those activities which are crucial to the business. Timely and relevant management accounting information is essential for the development, implementation and audit of strategy. It needs to be focused and kept simple.

10.7 SUMMARY

Companies have to explain the treatment of research and development expenditures in their accounts. These policies can have a significant effect on their reported performance and financial standing. The value of capitalized development costs, and intellectual property, such as patents and licences, must be stated in the balance sheet. Depreciation and other amounts written off have to be taken by the profit and loss account.

New technology, such as automation, has had a profound effect on the cost structure of manufacture. Direct labour costs have diminished in many industries and indirect costs increased. The apportionment of overheads to such a diminishing base leads to a distortion of cost structures and affects the pricing decisions based on them. The pressure of competition has increased the use of activity based and target costing systems as tools of analysis and cost reduction.

FURTHER READING

Accountancy Standards Committee (1989) Statement of Accounting Practice (SSAP 13) *Accountancy for Research and Development*, Institute of Chartered Accountants in England and Wales, London.

Batty, J. (1988) *Accounting for Research and Development*, Gower Publishing Co. Ltd, Aldershot. England.

Drury, J.C. (1992) *Management and Cost Accounting*, 3rd edn, Chapman & Hall, London.

Hopwood, A. (1974) *Accounting and Human Behaviour*, Accountancy Age Books, Haymarket Publishing Ltd, London.

Kaplan, R.S. (ed) (1990) *Measures for Manufacturing Excellence*, Harvard Business School Press, Boston.

CHAPTER 11

Appropriate technologies

11.1 INTRODUCTION

One of the ways in which industrialized countries have provided aid to the developing nations of the third world has been the provision of finance, equipment and advice to help with the building up of local industries. Over several decades of such assistance much experience has been gained from projects which, simply put, had failed. The reasons for failure have been well documented and one important quoted factor was the lack of appropriate education and technical training in the recipient country. Often the technology was too sophisticated; after the visiting specialists had left, the lack of the appropriate spectrum of human resources resulted in lower efficiencies and plant failures, the effects of which were compounded when the lack of foreign exchange precluded the purchase of replacement parts.

This experience fostered the development of 'intermediate technology', which concentrated on what was thought to be appropriate for the recipient population. As the name implies, it is neither primitive nor the latest state of the art available in an advanced industrial economy. Typically, the technology provided is simpler, the equipment is relatively low cost, unsophisticated but practical; it is easier to operate and to maintain and, wherever possible, local material, familiar to its users, is employed. According to Schumacher (1973), one of its best known exponents, intermediate technology is the technology level selected to match the realities of a local application, typically in a developing country. It is for the use of the masses but not for mass production.

11.2 APPROPRIATE TECHNOLOGIES FOR THE FIRM

It is argued that such an approach is not only valid for third world countries but for all nations whose population, for whatever reason, cannot make optimum use of the latest, sophisticated technologies. This is the case for 'appropriate technology'. It applies both to countries and to individual factories. For instance, one strength of Japanese industry is its systematic use of sound, safe and familiar technology. Sophistication for its own sake has no role; effective and reliable technologies matter. The point was illustrated by the UK Overseas Science and Technology Expert Mission to Japan on Intelligent Manufacturing Systems which noted in 1992 that Japanese CAD (computer-aided design) systems, seen in daily use, represented the effective application of established technology rather than high technology (Institute of Electrical Engineers, 1993). The systems were generally of a slightly lower level of sophistication than in the West.

A process may be similar but have different levels of mechanization and automation. Particularly, a multinational company may have similar processes with a range of levels from the relatively primitive to the highly sophisticated. This is very much a function of local skill levels and infrastructures.

For instance, a well-known electronic component firm has in one country a factory with highly sophisticated, fully automated assembly lines. According to the classification given in chapter 1 this is at technology level 5. In another country the same product is manufactured, using mixed automatic/manual assembly at level 3 or 4. The products are similar and to world class standards; their manufacture is based on appropriate technologies.

It is, of course, fundamental that the technologies a firm employs generate products with attributes and prices which are acceptable to customers. That does not mean that a company uses the most suitable technologies for its circumstances. There may well be some more appropriate technologies.

An appropriate technology is that technology which an organization, with given technical and management skills, educational levels, resources and cultural background, can apply to maximize the added value of its activities. A more advanced technology, which does not add further value would therefore be unjustified. Similarly, a less advanced technology would not yield sufficient added value. This applies to all business functions, not just to manufacturing operations.

Appropriateness is a function of social capability; i.e. the human resources within the company and the infrastructure on which the company depends. The assessment of what is the appropriate technology needs to be seen in dynamic terms. There is a short-run context and a long-term dimension. What is appropriate today may not be so at some future stage. It is therefore important that the matter is part of the periodic technology strategy review. Furthermore, an appropriate technology is not necessarily a competitive technology at world class level but it best satisfies the economic requirements of a particular community.

An appropriate technology reflects a human-centred (anthropocentric) production system, where the needs and capabilities of system members are considered in the system design. Where a third world factory is involved the inherent as well as the imposed cultures need to be considered. The local culture is bounded by indigenous languages which are often non-numeric. Much of the industrial training in third world countries is highly task specific and assumes only the rudiments of a basic education. There is also the problem of flexibility where staff have only been trained in narrow skills.

11.2.1 The choice of technology

The main determinants of a technology choice are economic, organizational, educational and cultural. A number of aspects are therefore relevant to the choice of an appropriate technology and these have been well summarized by Ramear (1979) in a case study of NV Philips. The following are among the more important:

1. the product/service specification required by the market,
2. the available managerial and technological know-how,
3. the scale of output required,
4. relative factor endowments and costs, e.g. wage levels,
5. the capabilities and adaptability of staff,
6. the spectrum and quality of available skills,
7. managerial risk aversion to production breakdowns,
8. the local infrastructure.

There are also aspects which influence the degree of utilization of the chosen technology. The following are the more important:

1. technology design and development,
2. technology cost and performance evaluation,
3. training staff for technology utilization,
4. technology maintenance and repair.

A firm needs to assess the state of the core technologies which it uses. Are they likely to advance quickly? Are there dominant, well-established designs and standards? Is a particular technology likely to be superseded? With a mature technology there is an opportunity cost of putting too much R&D into it when the same resources applied to an emerging technology can yield greater progress. The company can assess the degree of maturity by the knowledge diffusion within the industry and the extent of knowledge codification and standardization.

These are some of the criteria applied by potential users of a technology. As the following case example illustrates, corresponding evaluations are also made by the prospective suppliers.

11.2.2 An illustration of appropriate technology

Tidd (1991) quotes the case of assembly automation which in 1978 saw two significant developments. The first was the launching of the PUMA robot by Unimation in the USA. This was based on the state-of-the-art technology. In the same year a group of five Japanese firms began the SCARA (selective compliance robotic assembly arm) project and by 1981 when this robot came on the market the consortium had grown to 13 members, most of which were the prospective users of such equipment. Much of the design philosophy was based on USA research, which indicated that 60% of components for assembly were inserted from a single direction, while a further 20% were fed in from the opposite direction. An assembly robot would therefore only need limited dexterity but would require compliance to absorb positional errors. The result was a relatively simple design using two servo motors. Despite this the SCARA robot is claimed to be suitable for up to 80% of assembly tasks, but costs only about half of the more versatile PUMA.

Much of the automation in Japanese assembly lines is of a simple pick and place design and relies heavily on the line operators to keep it going. Consequently, much greater emphasis is placed on operator training in Japan than in the West. Operators supervising a group of, say, five to ten robot stations on a line are expected to clear machine blockages, monitor quality, carry out minor routine maintenance, as well as to obtain the required parts and carry out some assembly operations. Reliance on multi-skilled operators, linked to an emphasis on quality and design, has allowed Japanese manufacturers to adopt less complex technology and improve flexibility.

This example stresses that the choice of an appropriate technology requires analysis and the rigorous definition of manufacturing needs. Those needs are then matched to the technologies that use the available resources and build on the existing skills in the most effective manner. This applies both to developing and developed countries. There is no difference in principle; it is only a matter of degree. If anything, it will be more crucial in industrialized countries because there is no presumption of the need of such an approach. The belief that more advanced technology means greater efficiency must be evaluated by analysis. Advanced technology is introduced or 'transplanted' and if expectations are not fulfilled, then the 'perversity' of those involved is a tempting scapegoat, rather than the lack of an appropriate analysis in the first place.

A technology can be appraised in terms of levels of key attributes, for example with engineering components, the machining and measurement at a nanolevel of accuracy. Also, it can be seen in terms of other associated technologies, such as the comparison of a manual lathe with a computer-controlled lathe. It can also be assessed according to the required skills to use it.

11.2.3 An illustration of inappropriate technology

With a headline 'All hands to the production lines' – *The Times* on 29 July 1994 reported the fate of a robotic assembly line installed by IBM at its factory at Greenock, Scotland. When installed in 1986 the £6M computer monitor assembly line was claimed to be one of the most advanced robotic production lines in the world. In the autumn of 1993 the company removed the automated assembly tools and replaced them with operators. It now produces 700 monitors per shift with 50 assembly workers; an increase in output of over 25%.

The change led to better products and higher assembly productivity. Technical advances and specific customer requirements had widened the product range and increased the frequency of assembly process changes. Every such change required a line shutdown, the replacement of robot gripper mechanisms, and dimensional accuracy checks. In essence, the robot technology was based on flow production; the market had changed and required batch production. Human dexterity was more appropriate and flexible than advanced automation. 'Appropriate' technology had become inappropriate; the company responded accordingly.

11.2.4 The Japanese approach to flexible manufacturing technology

In his international comparisons of flexible manufacturing technology Tidd (1991) indicates that the Japanese suppliers and users of advanced manufacturing technology (AMT) have consistently developed and adopted less sophisticated systems than their counterparts in the West. On the other hand, American and European manufacturers have concentrated on more complex and expensive technology in an attempt to overcome organizational shortcomings and to match Japanese levels of productivity and quality. Having resolved the problems of improved quality, Japanese manufacturers have concentrated on increasing production flexibility with the application of relatively simple programmable automation. More product variants are produced per plant and product life cycles are shorter than in the West, despite the use of less sophisticated technology.

Several major Japanese machine tool builders have achieved overall cost leadership by producing standardised, lower performance, but significantly cheaper numerically controlled machine tools, which according to Jacobsson (1986) are about half the weight and a third of the cost of the machines produced by European and American manufacturers. The use of less powerful, but adequate drives has permitted lighter machine tool structures.

11.3 AN 'ABC ANALYSIS' OF TECHNOLOGIES

An informed visitor to various factories will note that most manufacturing companies use a wide range of technologies. However, there are great differences in their importance, relevance and levels. The mix varies with each business. The technologies can be classified and ABC analysis used to assess the value and relevance of a technology to a company at any given time. The criterion of value is competitive advantage. As we have already seen in Chapter 4 this can be due to cost leadership or differentiation. The latter is achieved primarily by equipment flexibility which allows a company to provide products of a greater variety or quality. ABC analysis is based on the Pareto principle, which indicates that in any statistical group a few members, the 'As', account for the bulk of the results. At the other end of the distribution there are many members, the 'Cs', whose combined impact is quite small, although not negligible. The third group, the 'Bs', are the intermediate category.

One of the uses of such a categorization is that it gives the company a ranking for the further development of and investment in a technology; i.e. to decide which technologies to take to a higher level. The very exercise requires a judgement about the relevance and promise of a technology. However, technologies are seldom isolated from each other; they often form constellations based on groups of scientific principles. Nevertheless, an ABC approach helps to focus the scanning of technological opportunities.

11.3.1 Core technologies

These are the 'As' and they are fundamental to the company's manufacturing operations. They are also known as the 'driving technologies' as they enable it to incorporate a product attribute, such as a fine surface finish, which gives its product a customer application advantage or yield a competitive production cost benefit. A grinding machine used by a precision engineering company would incorporate the core technology of precision grinding.

11.3.2 Contributory technologies

These are important, often generic technologies which are available to all the firms in an industrial sector. The technologies are associated with the core technologies, such as the hydraulic system that would be incorporated in the grinding machine. The user firm would not normally be concerned with the development of a more advanced hydraulic system but would take advantage of such progress when buying a 'design approach'. This is seen as a basic product or component design e.g. a jet engine or an aircraft undercarriage which will remain, in principle, unaltered for a number of years, but may still within its life cycle develop and achieve continuous improvement. The design integrates different technological innovations and encourages a high level of product standardization. It also provides a benchmark for competition in functional performance and costs.

11.3.3 Context technologies

These form part of the company's general operations and are usually embodied in proprietary equipment. They are mainly the 'Cs' and innovation activities are not normally undertaken by the user. For instance, a fork lift truck has a number of technologies embodied in it and the truck could be used in many different industries. The truck user's interest is primarily confined to the proper maintenance of the equipment, so as to reduce downtime and to avoid premature equipment failure.

11.4 A MACHINE TECHNOLOGY PORTFOLIO

This example is based on a computer-controlled machine tool (CNC), a three-axes machining centre which has become a standard precision engineering production unit. Figure 11.1 gives an outline view of the machining centre and Table 11.1 lists its main plant features and associated technological aspects. Its function is the programmed drilling, tapping, boring and milling of metal components.

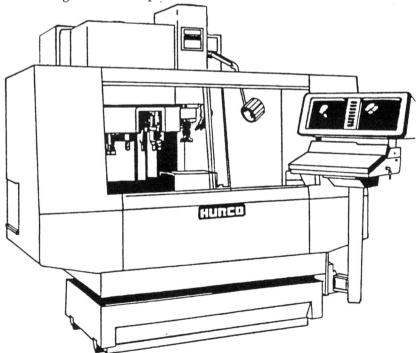

Figure 11.1 Outline view of a machining centre.

Table 11.1 Plant features of a machining centre

Plant feature	Key technological aspects
Structure	Framework stiffness, vibrations, alignments, precision movement of work table, slides.
Foundation	Rigidity, machine and external vibrations.
Power supply	Motors, starters, circuit breakers, power factor, AC spindle and servo drives, transformers, power and instrument circuits.
Drive	Range and variability of speeds and feeds, gears, clutches, spindles.
Controls	Operator dialogue, tool path, precision level, safety limits.
Tooling	Tool design, tool magazine, automatic tool change, tool wear compensation, tool presetting.
Work handling	Size, loading, clamping, chucks.
Utilities	Coolants, filtration, lubrication, swarf removal, heat exchangers for electric systems, hydraulic and pneumatic systems.
Programming	Data inputs and storage, modes of operation, inheritance displays, machine computations, standard functions and sub-programs, teach facilities, mistake correction, display.
Console	Diagnostic screen display, machine status, 'locations' screen display, graphics.

In this machining centre the core technologies, the 'As', for competitive advantage are concerned with the following plant features:

- controls,
- tooling,
- programming,
- console.

These give the machine its main functional capability and, other things being equal, are the technical basis for machine purchase.

The contributory technologies, the 'Bs' are relevant to:

- drive,
- work handling.

Their role is to some extent affected by the core technologies. For instance with work handling: if the loading of the machine is by automated pallet transfer then this becomes part of the overall control system of the machining centre. If it is manual, then a semi-skilled operator can load and set up the work piece.

The context technologies, the 'Cs', will be associated with:

- structure,
- foundation,
- power supply,
- utilities.

These are universal and as they are available to any user they carry no competitive advantage. This does not mean that they can be ignored. For instance, an ill-designed structure can cause vibrations in operation and generate defects. Provided that the design is to good professional standards and workmanship of proper quality, likely user requirements will be met.

It is possible to enumerate other technologies embedded in the design and construction of the equipment. Some are highly sophisticated and reflect the latest state-of-the art, such as the console diagnostic screen display, which is typically a level 4 technology as described in Chapter 1 section 1.10. Other technologies are conventional and have not changed materially for years, such as the machine coolant system which reflects a level 2 technology. With technological change some of the constituent technologies will alter. If the changes are minor and are confined to one or only a few technologies then the change can be regarded as incremental. This is particularly the case with standard, proprietary components, such as switches or valves. Figure 11.2 shows the concept of a technology portfolio related to levels of technology. A machine can be classified at intermediate technology level, say level 3, if its portfolio of technologies is limited and does not include any advanced technologies.

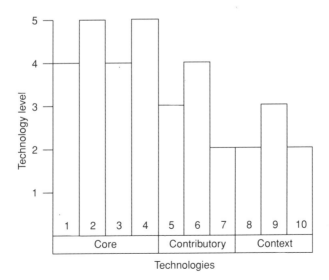

Figure 11.2 Typical portfolio and levels of technology for machining centre.

Some companies specify their general requirements for an equipment technology portfolio. Their needs can be related to the portfolios offered by suppliers and it is part of equipment evaluation to assess the closeness of such portfolios. A complete match often means custom-made equipment and the prospective user has to consider whether the cost of this is justified. It is part of contract negotiations to relate technology requirements to availability and price. Figure 11.3 indicates the integration of the supply and demand for appropriate technologies.

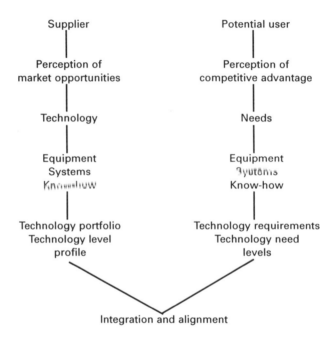

Figure 11.3 Appropriate technology: the integration of supply and demand.

11.4.1 The fusion of technologies

This example of the machining centre shows how a number of technologies can be fused within a machine tool. Some of these are generic technologies, such as the hydraulic system and the software packages. With the further development of one technology other constituent technologies sometimes become bottlenecks and the resultant concentration on their further development can bring significant advances to those technologies. Thus, innovation in one technology has a ratchet effect which leads to innovation in others. There is also cross-diffusion; once computer-controlled machine tools were established, the concept of computer-controlled coordinate measuring equipment was perceived and accomplished. The control principles of manufacture were applied to the control of precision measurement.

One important point about multi-technology machines is that these can be a mix of appropriate technologies. Although there will be interdependence, it does not follow that they all need to be at the state-of-the-art level. The economics of a particular technology have to be considered and that includes reliability and maintenance costs, the grade of labour required and the cost of incorporating a more advanced technology, including possible second order effects on other constituent technologies.

A telling example is the case of a computer-aided inspection system operated by a major UK aerospace company. In conjunction with a university research team it had developed a low-cost semi-automated intelligent inspection planning and computer-aided inspection (CAI) system. Full automation was not considered cost-effective. The developed equipment was a composite scheme for dimensional measurement. The manner in which operators used calipers was traditional; it reflected a level 2 technology. Digital data collection from calipers is, typically, a ten year old technology at level 3; whereas the integration of the data into a site data base is the current state-of-the-art, typically level 4.

11.5 THE OPERATIONAL SIMPLIFICATION OF ADVANCED TECHNOLOGY

While particularly in the initial stages of advanced technology, equipment and systems require greater skill and training to operate them, considerable attention has been given by the equipment suppliers to simplify their use. This has been one means to extend markets beyond the more sophisticated corporate customers and provides opportunities in the small/medium company sectors and in developing countries.

As an example, CNC control systems now have dialogue facilities, based on expert systems, which guide the operator in his work. Automatic programming is carried out by the machine in response to the moves of the operator. Much of the skill requirements, such as for tool adjustments are now carried out by automatic tool compensation subroutines. Errors are reduced by software travel limits and the programmable limitation of the working range. Again, the maintenance task requirements are reduced by interchangeable subsystems and automatic diagnostic facilities.

In essence, the skills and training requirements to operate and to maintain a CNC machine continue to be reduced. Of course, the capital outlay remains substantial but even here there is a long term decline in real costs. What is currently advanced technology will be 'intermediate technology' in years to come.

Another way of making advanced technology more appropriate is the use of expert systems in manufacturing. Such systems are designed to accept operator know-how. For instance, in a plastic moulding department pictorial arrays of mouldings are displayed on a screen, the operator chooses the moulding and the system gives expert guidance. Such systems are useful, for instance, for nightshift operation with skeleton staff.

Considerable attention has also been given in some industries, such as with the manufacture of electronic systems, to the development of automatic test equipment (ATE). The users will be:

- the operator who is a manual worker with modest skills and limited training;
- the ATE programmer, with skills in testing and diagnostic techniques;
- the maintenance technician who requires configuration and structural information.

11.6 FIRM-SPECIFIC KNOWLEDGE

It is characteristic of a technology that there is supplementary know-how which extends its application and usefulness. The know-how is often informal, i.e. uncodified, as well as acquired and transmitted by experience and example. Internal know-how can thus reinforce an appropriate technology within a company. Its ownership can be with various groups of staff. It can be semi-secret when operators do not divulge shortcut methods to supervision because of the fear of losing bonus or earnings. This occurs where shop experience is either disregarded or difficult to access by process planners.

The knowledge and cumulative experience required to constantly improve a production system normally originates at the shop floor level. It does not come from the higher echelons of management. In the Ono/Toyota system, rigid functional divisions are avoided; it is a bottom-up system based on the experience of the people within it and not on the stated capabilities of machinery. Relatively inexpensive reorganization of production processes and the mobilization of the knowledge and experience on the shop floor can, at times, yield greater dividends than major leaps in production automation.

11.7 THE TECHNOLOGICAL CAPABILITY OF A COMPANY

The level of technology and its appropriateness is affected by the technological capability of the company. This comprises the capability of necessary adaptation, sustained and effective operation, as well as the competence to maintain processes and equipment at a corresponding level. Outside sources of general knowledge and specific technology must be matched by in-house competence if they are to be effectively assimilated.

A leading UK electronics company, as part of an overall staff development policy, has a skill matrix inventory for its assembly workers. The qualifications, training and experience of each operator are on its database and a specific systems assembly and testing shop order can be matched within minutes with suitable assembly staff. The database is updated and reviewed at the annual appraisal interview. The company's policy is to have a flexible work force with every operator trained to a defined capability.

Where a company has such a skill inventory for operators, technicians and technologists, its capability to handle new technologies is readily defined. When necessary, an appropriate training and development programme can be tailored quickly to cover the gaps.

The introduction of new technologies emphasizes the need for job and application specific training. However, this can be over-emphasized as technologies evolve and change. It is also important for staff to see the business relevance of the emerging technologies.

11.7.1 'Automate or liquidate!'

This injunction has become fashionable. Like all slogans, it is simplistic and does not do justice to the complexity of technological change and the multiplicity of choices available to managers and technologists. Despite some widely publicized examples, such as the fully automated Japanese Fanuc machine tool factory, there is insufficient evidence to suggest that the majority of Japanese industrial plants have automated all their manufacturing operations. On the other hand, there have been cases of Western companies where the opposite has applied, where they have thrust themselves into automation schemes which proved such a disaster that their survival was threatened.

Appropriate technology is not the isolated adoption of automation, even with the expectation that an 'island' of automation will be the precursor of a comprehensive system of automation. There must be an effective balance between equipment, systems and staff which takes into account the capabilities and culture of employees. This does not mean the dilution of automation, but a change of culture and the development of staff to match the requirements of an automated system if the process economics warrant it. It may even be preferable to start on a greenfield site if the prospect of cultural change looks too daunting.

Advanced technology is by no means always an essential imperative to solve production problems. For instance, many of the flexibility gains which arise from computer-controlled operations can also be obtained from a 'low cost approach' which rationalizes set-up times and reduces costs, such as the Toyota SMED system advocated by Shingo (1985). (SMED stands for single-minute exchange of die; i.e. total machine set ups in under ten minutes.)

11.8 SUMMARY

An appropriate technology is that technology which allows an organization, with its given human resources,so maximize the added value of its activities. The choice of a technology and its level depends mainly on the product specification required for the market, the scale of output, wage levels and the capabilities of management and staff.

Most manufacturing plant incorporates a portfolio of technologies. These can be classified as core, contributory or context technologies and can be used at different levels. It is the appropriate level mix of the constituent technologies which maximizes added value. The operational simplification of advanced technology and the development of expert systems has reduced skills and training requirements.

FURTHER READING

Schumacher, E.F. (1973) *Small is Beautiful*, Blond & Briggs Ltd, London.

Shingo, S. (1985) *A Revolution in Manufacturing: The SMED System*, Productivity Press, Cambridge, MA.

Tidd, J. (1991) *Flexible Manufacturing Technology and International Competitiveness*, Pinter Publishers, London.

CHAPTER 12

The transfer of technology

12.1 INTRODUCTION

The first students of technology transfer were archaeologists: the subject matter is older than recorded history. The very terms: Bronze Age and Iron Age, reflect the nature of technology; the location of archaeological sites indicate its transfer. For us, the significant development is its growth this century into an explicit, major economic activity. The UK British Technology Group (BTG), a leading technology transfer company, has estimated the current value of world trade in intellectual property rights (IPR) to be of the order of £10 billion annually. Most of that involves technology transfer. The acquisition of technology from others is fundamental.

Quite separate and additional to the trade in IPR is technology transfer in the embodied form. It comes, typically, in the form of new machinery and equipment. Its most effective manner is people-embodied, such as with the support of commissioning engineers.

Technology is a tradeable commodity, often the subject of sophisticated contracts. There is supply and demand, but the market is very imperfect. As long as patents protect, a technology is private property and in many cases there is no wish by the owners to transfer technology which sharpens competition.

In 1992 the Royal Society, in a submission to the UK government, estimated that the UK conducted about 5% of global R&D. It strongly recommended that means were found for identifying and tapping into the best of the other 95% of global R&D. Here is the case for technology transfer. There is so much technology. It has become complex and expensive to develop. No firm or country could develop all the technologies it wishes to acquire and remain competitive.

Technology transfer has become part of business strategy. For instance, there is the firm with a well-established technological leadership in its field, but the further development of its core technologies is affected by technological developments outside its range of expertise. It decides to establish technological partnerships with the companies concerned or acquires the technology through a licence agreement. In some industrial sectors, technology transfer is now a major operational activity and a considerable amount of skill and experience has been developed by the specialist practitioners in this field. Technology transfer can be divided into the following categories:

1. Transfer from research establishments and universities to industrial companies. Often an industrial company sponsors or works in close association with the research unit concerned. Sometimes the academics concerned set up their own company to exploit a new technology.
2. Transfer between firms in advanced industrial countries. This assumes that both parties have technological, managerial and professional strength.
3. Transfer from an advanced industrial country to a developing country. In this case the recipient of the technology does not always have the scale of resources or experience of the supplier.

The operation of technology transfer is considered in this chapter at both the level of the firm and at macro-level. It is first concerned with the acquisition of technology and then with its marketing.

12.2 TECHNOLOGY TRANSFER

It is helpful here to develop the concept of technology transfer further. Technology transfer includes any transfer process which leads to knowledge being exploited away from its source. It encompasses transfers between sectors of the economy, such as from the universities to industry, between companies nationally or internationally; within companies, such as from a research centre to a manufacturing division and from one division of an international corporation to another in a different country.

Technology transfer relates both to the shift of proven techniques within an established technology and the introduction of a new technology. In the latter case the transfer can be combined with development. Technology transfer is an integrated process involving both the availability of technology and the needs of a customer.

The mechanism of transfer can be formal and explicit, such as by a licensing, know-how or information exchange agreement. It can also be achieved by the secondment or permanent transfer of staff, as well as by collaborative activities. In some cases mergers and acquisitions were regarded as the best way forward.

The transfer process, particularly for established technologies, is often carried out between firms, using their own specialist staff. Smaller companies and non-commercial institutions, such as universities, tend to use specialist technology transfer organizations. For instance, these organizations will assess promising ideas, potential inventions or new processes derived from academic research. If they are promising, the transfer organization will patent them, assist the development stage if necessary and, in due course, license them. The net licence income is shared, often on a 50/50 basis. Generally, the transfer organization has the licence negotiating experience and the contacts to place new inventions on a commercially attractive basis. This is particularly important where a technology is changing rapidly and prompt licensing and exploitation is needed.

12.2.1 The role of a technology transfer centre

Although much of technology transfer takes place within industries where companies know each other, there is also a role for national technology transfer centres. They can be particularly useful to small and medium companies who need advice and contacts anywhere in the world. A good example is the Technology Market Programme provided by the Centre for Technology Transfer (CTT) of the Singapore Institute of Standards and Industrial Research (SISIR). The programme contributes to the SISIR mission, the relevant part of which is:

> ... to lead Singapore industry towards greater international competitiveness through quality and industrial technology.

The objectives of the programme are to assist clients who are either searching for technology or who wish to supply it to others. Its main activities are at three levels:

1. Identification. Details are provided of suppliers and seekers of the required technology.
2. Assessment. This evaluates the technology on offer and the capability of the supplier.
3. Negotiation. The client is helped in the contract negotiations with an identified party.

The following network resources support this programme:

1. Computerized online databases. These include international online services for innovation and patents.
2. Technological datasheets and journals.
3. Technological transfer agencies. The international network includes the British Technology Group and the Japan Technomart Foundation.
4. Institutional contacts. These cover a wide range and diverse institutions, such as the Batelle Institute and the China Academy of Sciences.

The setting for these services is advanced information technology and the operations are carried out in a professional manner. The strategic intent is reflected by the integration of the programme with related activities such as:

1. The expert sourcing programme assists companies with the upgrade of their technologies by the identification of experts. The negotiation and the management of consultancies.
2. The R&D incubator programme furnishes subsidized laboratory and technological support facilities within SISIR.
3. The soft-start R&D programme provides companies with a cost-effective way to initiate industrial R&D projects.

This example shows how technology transfer can be integrated within an overall strategy and is clearly structured for effective implementation.

12.3 TYPES OF LICENCES

The licence to use a new technology is the most explicit illustration of technology transfer. It generally covers the use of patents owned by the licensor, together with such know-how as is appropriate. The following are the most important types of licence in law:

1. Exclusive: this gives the licensee the sole right to use the patent and deprives the patentee of the right of use. It is akin to the sale of a patent.
2. Sole: only the named licensee may use the patent apart from the owner.
3. Non-exclusive: this gives the licensee the right to use the patent, but its owner may grant similar licences to others.
4. License of right: where a patent is so endorsed, any person is entitled to a licence under the patent on such terms as may be laid down by the owner.

There are other forms of intellectual property which can be conveyed by a licence. The most important are process and operating know-how. There is also the special case of franchising a technology, such as with the bottling of cola and soft drinks.

The payment for a licence can be one lump sum, periodic payments, royalty payments assessed on production or sales. It also can take the form of payment by products. The precise form is a matter of contract and is primarily determined by legal, taxation or commercial aspects.

12.4 TECHNOLOGY TRANSFER AND BUSINESS STRATEGY

The approach by a company to technology transfer is part of its technology strategy. The parallel between make or buy, which is a decision for manufacturing management, is relevant here. As part of its strategy the company has to decide whether to develop or to buy a technology.

Similarly, it has to decide whether it will license its own technologies. These can be difficult decisions and there is evidence that many firms give insufficient attention to such problems.

For instance, a report by the UK National Economic Development Council (NEDC) (1989) on technology transfer concluded that companies do not view technology strategically, do not regard it as a long-term asset and do not think of technology transfer and accumulation as a long-term process. The Council recommended further study of the effectiveness of various technology transfer mechanisms.

Again, the British Technology Group (BTG) estimated that only 20–30% of a UK company's inventions were likely to be used within its business. Furthermore, relatively few companies licensed out their technology. Again, only a small proportion of UK firms were active in licensing technology from abroad although most R&D expenditures occurred outside the UK. Many did not appear to approach the acquisition of overseas technical information in a systematic fashion. Successful inward technology transfer required considerable long term commitment, often involving the build-up over time of a network of contacts with overseas sources.

A related issue is whether UK companies are sufficiently active in seeking technological alliances with overseas firms and whether there is sufficiently wide recognition that such alliances should complement, rather than substitute for, in-house expertise.

12.5 THE ACQUISITION OF NEW TECHNOLOGY

The acquisition of a new technology needs to be seen as an expression and implementation of corporate technology strategy. All the staff work and planning for such a strategy needs to be focused on the problems and manner of technology acquisition. Opportunism has to be integrated with a systematic approach and structured appraisal.

Much of a new technology comes in the embodied form, with new equipment. In that form it becomes available to the user, although there may be application development work and the need to establish operational skills. This is a matter of technology transfer where the equipment supplier is often an important actor. Plant commissioning, demonstration, operator training, instruction manuals are all instruments of technology transfer.

However, where a firm wants to harness a new technology which has to be developed, *ab initio*, the company has to make some basic decisions. The greater the resources required for the development of a new product or process technology, the more important will be the acquisition decision. Also, the company has to consider its R&D track record in developing new technologies to time. Its own R&D operational experience may not be in the required direction. Again, where strong competition is expected and market lead time is important the manner of technology acquisition can become critical. A number of alternatives are available and the company has to choose. As part of its business strategy it can do one or more of the following.

1. Develop the new technology itself. The company has to estimate the financial costs of the required R&D and the opportunity costs of that choice. Its impact on the direction of, and the commitment to, other research projects is also relevant. Furthermore, the company has to assess the suitability of its staff and equipment for the new project and may have to accept a considerable recruitment and training outlay. Among the risks it has to face are blocking patents. However, the developed technology can be customized to its precise requirements.

2. Buy the firm that has it. The investment can be substantial and great care is needed in the evaluation of the prospective acquisition. It is important, following the purchase, that the operations of that firm can be effectively integrated and that there is no undue loss of key staff. An illustration of this approach is a privatized UK water company which, as part of its strategy to become a world leader in the design, construction and operation of complete water treatment services, acquired five companies over a period of $3\frac{1}{2}$ years at a cost of about £200M with particular expertise and world markets in process instrumentation and equipment.

3. Enter into joint ventures. The costs are shared but so are the benefits of the new technology. Where the risks are high and the costs heavy, membership of a research consortium becomes a more attractive option. There is also the co-development of new products or processes, such as between a key supplier and a major customer. A particular form of current collaborative research is the UK Link Scheme, whereby government financial support is available to companies and universities for projects within defined areas of technology development.

4. Commission contract research. R&D contracts can be placed with research associations, universities or consultants. The company has to consider the costs and the nature of control of the project. There is the risk of know-how loss.

5. Obtain a licence for the use of the technology. This is essentially the purchase of access to proprietary technology. It can be anything from the right to use a particular patent to a complete package which includes know-how agreements, commissioning assistance for new plant and processes and the provision of updated designs and other technical information. Licensing here also includes the cross-licensing of new technologies where companies have formed technology alliances.

6. Education and training. Certain 'soft technologies', with a strong management dimension, such as JIT, quality circles, or Kaizen can be

acquired through training programmes but the underlying experience which makes these techniques more effective is often achieved through personal contacts between companies.

The options are, of course, not equally available. Getting a licence presumes a willing licensor. Much depends here on the technological competition between the two parties. Where licence agreements are between companies operating in different national markets this may not be a serious matter. The growth of global competition may change this.

12.5.1 The reasons for acquisition

There are a number of reasons why a company may prefer to obtain a technology where it has the opportunity to obtain a licence.

The first is the saving of research and development costs. This is not just a matter of finance. It also involves the marshalling of resources, particularly technologists and the time lag before they become fully effective. The purchase of a proven technology also avoids the risk of R&D failure. The acquisition of designs, specification sheets, process know-how and sample products simplifies the product launch and allows a quicker market entry. Scarce key staff can be assigned to other opportunities. Altogether, considerable savings in time and money can accrue and yield a faster return on the investment. Also, where the purchaser pays mainly through royalties on product sales this will defer his cash outflow.

Of course, the purchaser is unlikely to establish technological leadership with such an arrangement; although even here specialist marketing and application experience is often developed by him. But, as a matter of strategy, this may not be critical if the vendor of the technology is not a competitor in the market or country of the purchaser.

A related aspect that warrants attention is licensing conditions. For instance, in some cases there are 'grant-back' provisions which require the transfer of improvements, free of charge, to the licensor. If such information, in turn, becomes available to other licensees, who are competitors, this can impair the company's position.

12.5.2 The management of technology acquisition

Once the strategic decision has been made to acquire a technology from outside the company, the management of that acquisition becomes important. The UK Department of Trade and Industry (DTI), which is particularly concerned with the effective transfer of technology to British companies, recommends the following:

1. The role and management of technology within the company needs to be assessed, especially its capability of managing the transfer activity.

2. The allocation of appropriate staff to the transfer and application of the technology. The project manager must be at senior level and his colleagues need to have engineering application and change management skills.
3. The corporate objectives, capability and the technology transfer track record of the prospective transferor need to be considered. Effective technology transfer is often based on a longer term relationship.
4. Clear technical and contract specifications are essential. Because of the nature of the technology and its integration in intellectual property, the transfer constituents vary in type and character. Where the transfer is from a different culture, special attention has to be given to detail and the meaning of language.
5. Contract negotiations can be onerous. They require diplomatic skills and careful record keeping.
6. Because of the nature of its acquisition, transferred process technology needs to be handled with even more care than indigenous technological change. It is important that all affected company staff appreciate the nature and the reasons for the transfer.

12.6 THE MARKETING OF TECHNOLOGY

This section is concerned with the firm that has a technology to offer. Often, it has technology leadership in its field, but for various reasons it cannot or does not wish to exploit this in product form on a global basis.

As a technology is developed, becomes established and is well protected, its marketing becomes feasible. Such a possibility emerges when both the technology and its applications have been proven. By then a company is likely to have spent substantial amounts on R&D and also on protecting its inventions. In the first instance, therefore, it may wish to use its new technology as a temporary monopoly supplier in a growing market. When the products, incorporating the technology, become established in their own right then the value of the technology is increased. A particular concern for a smaller company is the release of a technology without the loss of its technology leadership or market share.

On the other hand, some companies have successfully developed a new technology which they cannot or do not wish to fully exploit on their own. Some of the reasons are lack of adequate finance, insufficient manufacturing facilities and experience, limited market strength or government restrictions in overseas markets. Where an industry is strongly competitive and the threats of infringement or circumvention of patents are serious then a firm, especially if it has only limited resources, may prefer licensing income to litigation expenses. For some markets, licensing a technology avoids import controls and customs duties which make it difficult for product sales. Porter (1985) also notes the competitive advantage of licensing a technology which is a prospective industrial standard.

The technology to be marketed need not be the mainstream technology of the company, but could have arisen as a by-product of research in other fields. The non-stick frying pan is one of the best known spin-offs of space research. A company may market this technology itself or through subsidiaries and affiliates.

It can also be company strategy not to get too involved in 'downstream' application work. The ownership of the new technology gives the opportunity of obtaining an income from licensing it to others. To maximize such income will require a licencing strategy akin to product marketing. It is also needs to be integrated in an overall business strategy.

A technology is usually marketed by licenses which give access to one or more of the following: patents, registered designs, know-how, equipment; product, process and test specifications, manuals as well as the services of expert staff. The licence may be on the basis of a lump sum royalties on future product sales or a combination of these. There is a distinction between start-up and follow-up licences. Start-up know-how is the information required to begin working a licenced product or process. A follow-up licence is mainly for accumulated process information and experience.

An important aspect for both partners of a licensing agreement is the treatment of technological change. The licensee usually expects the upgrading of the technology covered by the licence. In turn, the licensor is interested in application improvements. An exchange of technical information is a common feature of licensing agreements.

Ryan (1984) in his case illustration of the Pilkington float glass process, shows how important the licence income of a new technology can be. The process was licensed to many of the major glass producers all over the world. For Pilkington Brothers Ltd, the licence income amounted to over 50% of its profits before tax for a number of years. Significantly, in the evaluation of its investment in the float glass technology the marketing of the technology itself was just as important a factor as product sales.

There are cases where a company cannot cover all product markets, either because of the scale of the resources required or because of the risk of inviting powerful competitors to enter the field. If product penetration is an unrealistic business prospect then technology penetration may be the alternative. Apart from the licence income there are the benefits of standardizing on the basis of the most widely used technology. The risks of incompatibility and interface problems are reduced and the costs of adapting to other designs or key parameters are also avoided.

A fundamental aspect of licensing is the definition of the markets in which the licensee can operate. The licensor may wish to retain some markets for his own products, while the licensee, having often made major investments, will wish to have exclusive manufacturing and marketing rights in his territory. Frequently, there are also conditions on minimum levels of exploitation, such as the extent of incorporation of the technology in the licensee's product range or on the level of sales. The duration of a licence is generally linked to the patents which encapsulate the technology.

12.6.1 The marketing of know-how

As know-how relates more to practical knowledge and experience it is often person-specific and judgement-dependent. That makes it somewhat more difficult to identify its contents and boundaries and puts the onus on the careful drafting of contracts. Know-how has proprietary value and therefore is confidential. If it is subject to a licence agreement then the maintenance of confidentiality by the licensee becomes important. This may warrant the segregation of factory areas and the naming of staff who will have access to the know-how.

The provision of know-how generally includes technical assistance, such as the secondment of technical experts to the licensee's plant to commission the process, train staff and give general advice. The further development of know-how, due to innovations by either party, is an important contractual matter. The duration of the agreement is often governed by the expectations of technical change or the diffusion of the know-how which, once it becomes general knowledge, loses its proprietary value.

There is an art to the transfer of know-how and the related skills are not freely available. Know-how is best explained and assimilated within a context of background information. The staff who supply it must be able to adapt know-how to different environments and resources. Their communication skills must include an awareness of whether they have been properly understood. To allow for the process of training and assimilation, especially with developing countries, it is often better to make technology transfer a gradual process.

12.6.2 The transfer of technologies within multinationals

A good account of the issues of technology transfer is given by Veldhuis (1979) in a case study of Unilever practice. Transfer does not just consist of a set of separate technologies, but also includes a synthesis of the relevant management skills and know-how. Otherwise there is the risk that the separation of technologies into specific parts inhibits the ability to tap the full know-how of the originating division. An operational technology then tends to become separated from the background technologies which make it understandable.

It is assumed that the technology supplier has the complete know-how relevant to his own environment and that either he or the recipient has the

R&D capability to adapt it to the new setting. The recipient company must have, or is able to develop, the management and technical skills to accept the technology in full. The supplier must be prepared to give technical and management support for a period after the transfer, preferably through personal contacts, visits, secondments or training courses.

12.6.3 Technology transfer to a developing country

Where the transfer of technology is to a firm in a developing country the supplier needs to pay much more attention to the economic, social, political and cultural environment of the buyer. The country has to have a suitable industrial infrastructure so that it can utilize the imported technology. This requires a level of national technological and human resources to make effective use of the transfer.

At the level of the firm, the normal commercial safeguards are required, especially if the technology transfer requires a substantial investment in equipment. The supplier of the technology also has to satisfy himself that the buyer can make and maintain proper use of the new technology. A licence creates a partnership and it is to the advantage of the supplier to know how his technology will be used and what skills are applied.

The label of 'technology transplant' is applied to the international transfer of technology without regard to the cultural and organizational contexts of the recipient company. It is not confined to developing countries and like the transplanting of a grown tree, it has its risks. Apart from inefficient utilization, such a transfer can generate antagonism to its application. Where the transfer is expected to cause redundancies then industrial relations aspects can become a significant obstacle to the use of the imported technology.

12.7 TECHNOLOGY TRANSFER AS A NATIONAL CONCERN

Since 1945 there has been a major change in the world-wide location of industry. This global shift has been well depicted by Dicken (1986). New industrial nations, such as Korea and Taiwan, have developed and Japan, with its rapidly growing global industrial base, is assuming a much greater leadership role in advanced technology. All this has been achieved by a mixture of motivation, strategy, imitation and innovation. The reward has been rapid economic growth and a higher standard of living. Such success invites copying and many countries want to improve their economic performance by the greater use of advanced technology. They want to catch up. One way to do this is to encourage an inward flow of technology.

12.7.1 The strategy of transfer

One of the best examples of national strategy applied to inward technology transfer is the industrial revival of Japan after 1945. Its most telling feature in the 1950s and 1960s was the systematic purchase of technology from the West. The *Second Report of the House of Commons Select Committee on Science and Technology*, (Japan Sub-Committee) Session 1977–78, (HC682–I) is a fertile document on the inward flow of technology into Japan during the period 1955–75. It reports, for instance, a survey by the Japanese Science and Technology Agency, based on the replies of 645 corporations. This indicated that 42.3% had introduced technologies from overseas. Subdividing this figure, the use of these technologies made by the firms is significant:

19.4% → combined with home developed technology;
18.1% → improved in Japan;
4.8%   → used unmodified.

Imported technology has been well digested and often much improved. A generation later the benefits of the strategy are clear.1 In addition to the huge annual trade surpluses, there is now a fast-growing export balance for technology. Japanese technologies transferred to developing countries are 'general technology', i.e. they are common knowledge in the industrial countries. An example of this is the Japanese–Malaysian partnership in the development of the Proton range of cars. High level technology, such as generated by the pharmaceutical or electronics industries is licensed to companies in advanced industrial countries.

12.7.2 Taxation and technology transfer

The systematic acquisition of technology from abroad can be a heavy drain on foreign exchange. Many countries cannot afford such a pressure on their currency reserves and tax and restrict corresponding transfer payments. They have been particularly anxious to curtail the opportunities available to the multinationals for flexible transfer pricing. The most common fiscal methods are:

1. A withholding tax on the payment of licence royalties to a foreign licensor. Although most double taxation treaties aim to eliminate this tax on royalties, there are some countries, including Japan, that do not grant a zero withholding tax rate on licence royalties in their treaties.
2. A tax on the transfer of technology to foreign affiliates.
3. Restrictions on inter-company payments for technology.
4. Transfer pricing adjustments for technology transfer.

12.7.3 Assistance for developing countries

It is common for many aid programmes to be furnished by industrial exports to develop the infrastructure and industries of third world countries. The problems associated with 'aid technology' have been well documented and have given rise to the concepts and policies of intermediate and appropriate technology. These are not developed here but interested readers are referred to Schumacher (1973), perhaps the best known advocate of intermediate technology.

However, there are some aspects of technology pointed out by Edosomwam (1989) which are pertinent here. In essence, technology transfer to developing countries needs its own infrastructure. The following is required:

- the training of transfer agents and users of technology;
- the formation of technology transfer centres and maintenance centres;
- the establishment of training centres for existing and emerging technologies;
- procedures for establishing real needs; for the screening, justifying and modifying of technologies and diffusing the appropriate technologies to the required sectors of the economy.

12.8 SUMMARY

It is often cheaper and more convenient for a company to buy a proven technology rather than to develop it. The transfer of technology is a fast-growing business and has established itself as a component of corporate strategy. It is one important way to acquire a new technology. Its growth has been fostered by specialist technology transfer centres which identify, assess and negotiate with the owners of the chosen technology. Technology transfer is primarily by means of a licence. This can be exclusive, sole, non-exclusive or a licence of right.

New technology flows from universities and research establishments to industrial companies. Technology and associated know-how is acquired and marketed by corporations and transferred between countries. Technology transfer is becoming increasingly a national concern. The search for new technology has been facilitated by access to world-wide databases.

FURTHER READING

Dicken, P. (1986) *Global Shift: Industrial Change in a Turbulent World*, Harper & Row Ltd, London.

Edosomwam, J.A. (1989) *Integrating Innovation and Technology Management*, John Wiley & Sons, New York.

Hearn, P. (1981) *The Business of Industrial Licensing*, Gower Publishing Company, Aldershot.

Ryan, C.G. (1984) *The Marketing of Technology*, Peter Peregrinus Ltd, London.

Schumacher, E.F. (1973) *Small is Beautiful*, Blond & Briggs Ltd, London.

CHAPTER 13

The management of improvement

13.1 INTRODUCTION

This chapter is concerned with improved performance and the more effective use of given technologies. It emphasizes the role of management practices and company culture to attain this. It stresses learning and putting to use what has been learnt.

Many industrialists, after visits to Japanese firms, have commented on the modest differences in the levels of technology used by the average Japanese manufacturing company and that of their Western equivalent. A research investigation into the comparative performance of Japanese and UK automotive component companies by Andersen Consulting et al., (1992) indicated that Japanese world class plants were more automated and had higher production volumes. Nevertheless, they estimated that these two factors together accounted for less than 20% of the performance gap. The observed differences in productivity and product quality were due to management practices, the particular emphasis on continuous improvement, 'lean' manufacture and close cooperation with suppliers.

The introduction of major technological changes is not a daily occurrence. There are, for a number of reasons, intervals between the big steps of strategic innovation. However, there can be a flow of continuous incremental change: an upgrading, improvement, a search for better methods; greater efficiency, higher yield. We are now concerned with the more effective use of given technologies. Referring to Chapter 1, Fig. 1.1, the improvements will be primarily in the refinement of techniques, the extension of know-how and the better use of equipment. It also illustrates the attainment of best practice, as shown on Fig. 3.5. The maximum benefit is extracted from a given set of technologies. Much of such progress is unspectacular and escapes the outsider's attention, but its cumulative effects can, at times, be decisive. Another feature of such incremental change is its relatively low cost. Much of its input is ingenuity.

Chapter 7 discussed the role of change programmes which relate to the introduction of new technology. The management of improvement also calls for change if the organization and culture within a firm is an obstacle to improvement. Often this requires complete changes of attitude without the stimulus andopportunity of new plant and technologies. It is a challenge to management within a given physical context.

The management of improvement has two components. The first is a continuous programme of improvement and/or cost reduction without a major change of technology or facilities. This relies heavily on the motivation of workers and supervision. The second is a change programme which involves the rearrangement of facilities and organization. While there is still emphasis on work force motivation much of the executive responsibility rests with a task force of technical and managerial staff.

13.2 DIFFERENCES WITHIN A COMPANY

While many comparative studies have looked at companies in the same industry and broadly with similar technologies, it is enlightening to compare similar manufacturing plants within the same corporation.

Chew *et al.* (1990), in their study of US manufacturing plant productivity levels, note a wide range of performance levels between plants of the same company, although they used similar technologies and manufactured similar products. Even allowing for differences in the age, size and location of the plants, differences of a ratio of 2:1 could be observed. While there are several factors which could account for such differences, they note that the information flow between plants is not necessarily standardized or routine. Much of the information about incremental process innovations is qualitative and not quantitative and can, at times, be difficult to communicate precisely.

The persistence of significant productivity variations within a group was attributed to the values and beliefs that guide local managers and those responsible for the information networks, measurement systems, incentives and staff functions. Their account shows the opportunities which exist within a corporation to improve its operations.

An example of taking such an opportunity is the strategy of the Ford Motor Company in Europe, with its 'co-location' of engineers from Britain to Germany and *vice versa*. Those employed on particular projects work side by side, irrespective of national affiliation. Such an interchange of engineers between Dunton and Cologne was seen to improve the efficiency of vehicle development.

13.3 CORPORATE CULTURE

If the objective is continuous and preferably frictionless improvement, how does one go about this? There are a number of things that need to be done. To start with the basic requirements: an appropriate culture of the firm, management style and reward system are important requisites for improvement. Culture is seen as a function of values, expectations and beliefs. Twiss and Goodridge (1989) interpret it as the body of norms, assumptions, conceptions of value and attitudes which affect the actions of employees at all levels. It forms the character of the firm and the problem-solving behaviour of its staff.

Denison (1990) considers that the following four attributes determine the culture of an enterprise:

1. Involvement: is concerned with the employee's degree of participation in the activities of the company. It can be assessed by the responsibilities staff undertake and the sense of ownership they feel.
2. Consistency: is an indication of the coherence and stability of the ruling values and beliefs throughout all functions and levels of the business.
3. Adaptability: concerns the ability of the organization to respond to changes in its environment. This is expressed in terms of external and internal customer requirements.
4. Mission: is seen as the shared definition of the purpose and the function of the business and its members.

If, in order to achieve a climate for continuous improvement, the management of a company wants to attain a change of culture it needs to:

1. translate the new values into its own behaviour;
2. communicate the new values to all levels of the organization;
3. provide training to support the new values where needed;
4. shift power towards people whose skills, values and behaviour support the new culture;
5. identify, where possible, high-impact systems and use them to affect staff behaviour;
6. reward the desired behaviour;
7. have patience to allow the new culture to take hold.

Basically, what is involved is a change in the states of mind of individuals and groups. New patterns of interest, behaviour and habits are called for; the inertia and 'comfort' of existing routines has to be overcome. This is no mean challenge for a management and will require time, effort and resources. Many companies have become aware of this when initiating total quality programmes. Much can be gained from symbolic actions, such as product demonstrations, staff clinics, indeed anything which makes the mission vivid in the minds of staff.

13.4 CONTINUOUS IMPROVEMENT

The only source of lasting competitive advantage is a company's ability to develop and release the full potential of all its people more effectively than itscompetitors. Continuous improvement is a facet of competitive strategy and requires a framework for its realization. This starts with an 'improvement target' which, in turn is the basis of the 'improvement programme'. Its nature can vary from a component of company strategy, developed and implemented by key staff to guidelines for operational management. It can contain removal schedules for bottlenecks, which are constraints to productivity, or 'quick hit' lists which promise high returns for staff efforts. Strategic changes can be brought about by small steps. These must be manageable within company resources and allow time to prepare and train staff. A realistic rate of assimilation of a new technology needs to be established. There is also a need for an 'improvement budget' to relate activity costs to improvement benefits.

Continuous improvement at operational level is more than labour saving; it encompasses all advances which yield reduced costs and/or added value. To achieve this in systems terms there needs to be:

- an encapsulation of current knowledge;
- adaptation to changing conditions;
- a method to capture and incorporate new ideas.

Sound production engineering is often a low-cost route to the improvement of manufacturing performance, particularly within the framework of a cost reduction programme.This does not necessarily call for heavy investment and high technology solutions but for industrial engineering thathas achieved the reliability and flexibility of equipment.

The management of improvement also includes suppliers and the greater the purchased proportion of product value, the more important become joint programmes of continuous improvement and cost reduction. Especially for the major first-tier suppliers there needs to be a framework of joint development which capturesideas and exploits common innovation opportunities. There often is a Pareto distribution where, say, 20% of the suppliers acocunt for about 80% of purchase value. These can be targeted together with any critical items from the remainder.

An audit system and a set of evaluation techniques are an important feature of a continuous improvement programme so that improvements can be assessed, measured and costed.

13.4.1 An organizational setting for improvement

Because of the speed of technological change and the diversity of new technologies, there is a greater need to delegate more of the decision making to middle management and supervision. This is intended to lead to faster decision making, within agreed policies, and to make fuller use of specialists at operational level. Managers and supervisors really have to know the details of their operations and be more familiar with the overall business performance and their contribution to it.

One important objective for such an organizational setting is the elimination of 'technological slack'. This is interpreted as the state-of-the-art technology which is known to staff but for various reasons is not applied to production processes. Often the reasons are failure to deploy, lack of human resources for the staff work involved or preoccupation with other problems.

An appropriate organizational climate calls for as much management attention to information flow as the physical material flows in a factory. Communications between units and functions don't just happen and if they do, they do not necessarily support the goals and plans of the business. An information system needs to be planned; there neeeds to be a specification for it and it needs to be updated in the light of actual experience.

13.4.2 The level of innovation impetus

Whatever the precise source of stimulus, major technological innovations are introduced and managed as part of a top-down process. This is particularly the case where there is high risk, substantial investment and a long lead time with a new technology. Too much is at stake for top management not to take charge of such developments.

The bottom-up process or bubble-up approach is where ideas flow freely from employees at all levels within the company. In most cases successful ideas lead to simple, low-cost innovations which can be introduced relatively quickly. The cumulative effect of such ideas can, of course, have a significant effect on the competitiveness of the company. Most ideas which come up this way are fragile things; they need chaperons to survive even before evaluation and chaperons to sustain and to develop them once their promise is apparent. The infant mortaility of new ideas is high and it is the chaperon's task to ensure that good ideas do not die because of neglect. Some companies with a strong commitment to the bubble-up approach have appointed managers whose main role is to encourage, detect and to sponsor the development of new ideas.

The top-down and bottom-up approaches are not exclusive. Their integration within a 'U-turn' style of decision making is practised by some Japanese companies.

13.5 THE LEARNING ORGANIZATION

A learning organization is one which can adapt itself readily to changing

business needs and to the context in which it operates, while simultaneously continuing to improve its own internal operations. The speed of
learning is a vital factor in what Kodama (1991) describes so vividly as the
transformation of a producing into a thinking organization.

While learning is ultimately an individual activity, Cyert and March
(1963) and others have stressed the importance of the organizational
environment in which such learning takes place. The acquisition and
assimilation of new information, which are components of the learning
process, are affected by the training and experience of those within the
firm. There is the interaction of hope and expectation which gives a bias
to the search process. Organizational learning is a function of goals, attention and search rules. It is expressed in terms of routines which, in turn,
are the product of previous experience and the culture of the company.
According to Levitt and March (1988) such routines not only include
rules, roles, habits and conventions, but also structures, strategies and
technologies in which beliefs are embedded.

It is important, therefore, that such a framework of routines is
harnessed to discussions as to what is competitive performance, so that
minds are focused on this subject. It has often been noted that organizational learning is most effective if it is parcelled into small steps, solving
one problem at a time. There needs to be personal attention and that is a
scarce resource. There also nees to be the will to change and that means a
break with established patterns and carries elements of risk. Change,
therefore, is best structured in the form of manageable and comprehens-
ible projects. Once the decision has been made to proceed, it is important
that the instructions are achievable. If the chances of success fall below,
say, 95% a credibility gap will develop. For any particular group of staff:
one step at a time; concentration on one problem at a time is often the best
approach. A multiplicity of technical advances can unsettle a work force.

13.5.1 The culture of continuous learning

The changes in engineering production processes, due to the diffusion of
automation systems, have emphasized the new skill spectrum required
by manufacturing staff. Many of the skills that have to be applied by an
operator or technician derive from a range of traditional craft skills, such
as fitting, welding and wiring. Continuous and very specific training is
now required for the new portfolio of skill requirements.

These training programmes have to consider the psychological needs
of personnel for self esteem and personal growth. For many work staff
going on training courses has a flavour of being remedial, of 'going back
to school'. Such an outlook is somewhat passive and loses the initiative
dimension of continuous improvement. To change this outlook, as part of
its employee assisted learning scheme, a major UK car manufacturer
gives its workers up to £100 a year each to spend on any training course

they like, from foreign languages to computing, irrespective whether this is relevant to their work task or not, just to get them accustomed to the routine of continuous learning and self improvement. The willingness to learn is considered a crucial state of mind.

13.5.2 Limits to incremental change

An interesting concept of 'morphostasis', has been put forward by Tranfield and Smith (1990). The basic operational order is preserved and the process of incremental change is that of adaptation. Disturbances are regarded as 'external noise' to an operational system requiring minor adjustment. Where the requirements of change become substantial and approach a step function there is a need for transformation which they call 'morphogenesis'. The system can no longer take the changes in its stride and management needs to have a vision building approach coupled with strong leadership.

They demonstrate the changes and their implications in the form of a cube with technology, organization and business along the three axes, each with a 'high' and a 'low' point scale. Business is defined as the competitive position of the firm and organization includes job design linked to human factors.

13.6 THE FRAMEWORK FOR INNOVATION

The world is full of ideas, but good ones are in short supply. As we have already seen, innovation, in the first place, is the application of inventions to industrial and commercial purposes, typically to products and processes. This is the responsibility of the company technical function. But apart from inventions there is a whole range of non-proprietary ideas which emanate from the day-to-day operations of a business. How are good ideas, which are essentially derived from signals and stimuli, picked up in the first place? By whom? And how can they be assessed and harnessed if found promising? This is not just a matter for the firm but also applies to every individual. There is a strong suspicion that most of us are relatively inefficient in the capture and utilization of ideas.

In the context of the firm there are the additional impacts of the prevailing culture, derived from the existing socio-technical systems and group dynamics impinging on individuals. There is the interface between the group of people which constitute the firm and the environment in which they work and live.

We have already noted in Chapter 6 the various sources of ideas which can contribute to the formation of a R&D programme. The management of improvement is concerned with ideas for every activity, especially manufacturing operations. Relatively few ideas emerge from a vacuum.

An idea is usually the product of a signal, which is an information transfer, and a stimulus, which is a function of a state of mind. Ideas, of course, are generated within the normal day-to-day dialogues between staff. They are also the product of various brainstorming sessions. These two sources are part of, or are easily integrated within, a standard information system. They have a chance of management attention.

The classic example of structured idea gathering is, of course, the company suggestion scheme. Some companies have run these successfully for many years. Success here depends very much on the style of management and the culture of the business. The role of suggestion schemes is to stimulate awareness and thought especially among operators, who because of their role become aware of shop floor improvement opportunities which had not been envisaged by functional staff.

An environment in which there are many stimuli – not necessarily comfortable ones – is more suitable for the gathering of ideas. Coupled to this must be a level of awareness; perception is a gift which is not uniformly distributed but can be improved by motivation and training. Here then is a first responsibility of management: to increase the alertness of staff in all functions and at all levels. In the context of technology there are the opportunities of conferences and exhibitions, learned and trade journals. Less specific, but of value, is personal contact, such as participation by a manager in the affairs of a trade association or the perceptiveness of a sales engineer when acquainted with the problems of a customer.

The gathering and prompt assessment of ideas is the second management responsibility within the framework for innovation. The evaluation of an idea is a filtering process which can be anything from a quick look to a hierarchy of carefully framed assessments. An effective management has a structure for its filtering activities. Obviously, the idea is first looked at in terms of implications, costs and promise for the firm. Filtering is an energy consuming task and at the first point of review little energy may be spent; the obvious non-runners are removed. However, there is not always agreement on this and that is one reason why, with a suggestion scheme, an idea is assessed from a number of different views. The important thing is to have an effective catchment process which forwards promising ideas to the structured review systems of the firm.

It is important, of course, to manage ideas so as to avoid the alienation that may be caused by an ill-judged manner of rejection. Many suggestion schemes are careful about this and reward ideas even if they are not accepted.

13.7 CONTINUOUS PRODUCT INNOVATION

Porter (1990) stresses the active search for pressure and challenge. To sell to the most sophisticated and demanding buyers and channels may cause stress, but this is also an improvement opportunity. Buyers with the most difficult needs provide product application opportunities and yield inputs for development programmes. The satisfaction of regulatory requirements and product standards provides a benchmark which all suppliers need to meet. Targets which exceed such hurdles are the basis for product and process improvement programmes.

Incremental innovation is the key to shorter development cycle times. An effective way to progress quickly is to take many small steps frequently, within a broad framework of a product development strategy. It is important, however, in fast moving industries to coordinate changes and to integrate them promptly into product updates. Many Japanese companies introduce incremental product improvements on a regular basis during the product life. There is no big collection of improvements saved up for the next model. This evolutionary approach reduces the risks in the market place and also improves the prospects of smooth production start-ups. The direction of product evolution is also considered at the launch of the new product. For instance, the initial definition of a range of key dimensions can already be taken into account at the initial design stage, such as with equipment and fixtures. They don't have to be redesigned midway in the product life.

The Japanese emphasis on simplicity of manufacture is linked to step-by-step product innovation. They start by automating simple tasks, not the most complex. Equipment is often built in-house and introduced with a high level of shop floor involvement. When proven in operation it becomes standard for all similar functions within the firm.

13.8 KEYS TO IMPROVEMENT

13.8.1 Kaizen

According to Imai (1990) the essence of the Japanese term 'Kaizen' is improvement. While it is perhaps best known in the context of industrial work improvement, it applies also to personal life outside employment. In some ways it is a philosophy of life.

(a) Recognition of problems

Kaizen starts with the recognition that all companies have problems and that it is an important aspect of corporate culture that such problems can be freely discussed. Problems can be within and across functions, particularly with the development and launching of new products. Kaizen relates improvement to increased customer satisfaction – every activity is related to and judged by this. Customer satisfaction is not easy to define but can be assessed in terms of quality, cost and availability of the product in the right place at the right time. These attributes can be summated within the term 'value'. It is, however, necessary in some cases for management to establish priorities between these different constituents of value.

The recognition of a problem is the first step towards improvement. Problem awareness is therefore important and staff need to learn how to identify and to define problems.

(b) Scope of Kaizen

Kaizen is seen by some as an umbrella term which includes a number of management techniques and practices that are already well established in their own right, such as total quality control (TQC), quality circles, the Kanban and the 'just-in-time' (JIT) methods of production control. It also encompasses cooperative labour-management relations and work place discipline. The management and effectiveness improvement of the supply chain is also included.

In the Kaizen view the management function contains both a maintenance and an improvement role. The first role is concerned with the maintenance of current technological, managerial and operating standards. Further improvements depend on effective standards in the first place. According to Imai an effective standard has the following characteristics:

- individual acceptance and responsibility;
- transmittal of individual experience to the next generation of workers;
- transmittal of individual experience and know-how to the organization;
- accumulation of experience (particularly failures) within the organization;
- deployment of know-how from one workshop to another;
- discipline, i.e. everybody works to it.

Standards are the achievement of previous improvement programmes and need to be followed – there can be no drifting or slippages from these. It is, of course, everybody's task to maintain them but in addition much stress is laid on the responsibility of management to secure improvements of these standards. The higher the level of management the greater is this responsibility. Some Japanese managers maintain that they should spend at least 50% of their time on improvement.

(c) The next process is the customer

While the basic purpose of Kaizen for a company is to increase customer satisfaction in the market place, customer satisfaction within the firm is also expected in the next department or process in the flow of production.

Customer satisfaction in the next downstream department helps that department to add effectively to the product value which, in turn, yields satisfaction in the market place. The next department is not only within the production flow, but in the cross-functional sense includes manufacturing as the customer of the design activity.

(d) The quality of people

Within the framework of Kaizen and TQC the quality of people is of primary concern. A business can be viewed to consist of hardware: buildings, equipment etc., software: systems, procedures, computer packages etc., and 'humanware'. The latter, not being a balance sheet asset, is too often underestimated.

For many manufacturing operations, apart from a basic minimum level, the precise portfolio of qualifications and skills a person brings to his work is less important than the ability and willingness to learn continuously and to adapt. As many experienced managers know, attitude in an important accot.

(e) The application of Kaizen outside Japan

While the Kaizen philosophy reflects a Japanese way of life, the success of its application has prompted some Western companies to emulate it. It was felt that the Japanese experience could be assimilated and applied, if a serious effort were made. For instance, *The Times* of 24 August 1992 reported on its adoption by the Ford Bridgend Engine Plant. Its manager was confident that Japanese productivity levels were within reach. The source of Bridgend's competitiveness was no less than the collapse of the age-old antagonism between capital and labour. The plant convenor representing 1100 members of the Transport and General Workers Union at Bridgend stated that as far as the management/union team was concerned there was just one objective, and that was the prosperity of the plant. In a bold move, Ford adopted a policy of disclosing all business information to the work force. At Bridgend this was followed by the creation of working groups, each comprising 10 to 12 people with a mix of skills. Each group is responsible for part of the production process, answerable for quality and productivity. The process was accompanied by training programmes, and the establishment of joint management and union 'conferences' to oversee the plant business plan and product quality.

In productivity terms, the results were self-evident. The downtime of the engine block line was reduced from 45 to 19% and for the cylinder line from 55 to 26%. The plant manager concluded that you did not need a lot of investment for the improvement of performance, if people had the skills and were allowed to make decisions.

13.8.2 The concept of 'Poka-Yoke'

This concept is especially concerned with the improvement of product quality by preventing defects. Poka-yoke is based on the Japanese words 'yokeru' – to avoid and 'poka' – inadvertent errors. Its objective is the complete eradication of all manufacturing defects. While it reflects an important underlying philosophy of management, it also furnishes a significant facility for incremental change. It concentrates on the automation of quality assurance and brings with it detailed technical changes to specific manufacturing operations. Often this is achieved by surprisingly simple design features, such as the modification of a screw head to avoid screwdriver slippage that scratched cassette covers on assembly. Sound production engineering is enhanced by the use of sensors, limit switches and other electronic devices (autonomation). The main advocates of poka-yoke, Nikkan Kogyo Shimbum Ltd, (1988), quote 240 examples chosen from a wide range of manufacturing operations, which include machining and fitting work, electronic and mechanical assembly, painting, printing, sewing and packing. All the examples are given in the same format with clear visual presentations (even comic sketches if these help) of the processes concerned and in a manner which facilitates the adoption of new ideas.

13.9 PARTICIPATION IN CHANGE

Section 7.12 referred to three types of change management. This section is concerned with the third of these, where change is incremental and adaptive.

Apart from the behavioural aspect of being party to a change there is a need to tap the know-how and specific process experience obtained by production operators over a period of time. In many Japanese companies JIT and total quality procedures are nurtured on the shop floor. Workers and line managers are the focal points. Technical staff don't have the time to continuously assimilate minute details of operation. Participation gives specialist staff listening opportunities.

13.9.1 Gaining from conflict

Change often engenders conflict. This is not only because of the perceived gains or losses by various interest groups within the company but also because of different approaches to problems and related value judgements. It is important to have an effective way of resolving conflicts and to respond to the challenge of achieving consensus. The analysis and integration of seemiongly opposing views is creative and can often result in better solutions. Also, when consensus is achieved all those concerned will commit themselves to the agreed way forward. The investment in time to achieve consensus is more than repaid by speedier implementation. This is one of the benefits of the Japanese 'ringgi' form of group decision taking.

13.10 PLANNING FOR CHANGE

As mentioned in the introduction to this chapter, the second component of an improvement programme is concerned with larger changes than incremental improvements. The project content is much larger and the planning tasks become crucial. According to Lupton (1965) there are two objectives which management has to consider when planning for change. The first is to reduce the amount of disturbance during the period of change. The second is to move as swiftly as possible to a new and stable level of performance. He suggests that planning for change needs to include the following:

1. the listing of all alternatives of implementing the change together with a time schedule for each;
2. the identification of all sections affected directly or indirectly by the proposed change;
3. the assessment of the likely reactions of these groups; both in general and in respect of specific matters, such as wages, promotion prospects, etc;
4. an estimate of the overall acceptability of the change and the various approaches to implementation.

13.10.1 Change programmes

One of the greatest challenges facing manufacturing industry is how to manage the pace of change that is required. A change programme is more than a quest for greater efficiency and cost reduction. It must be based on strategic objectives. As a company's markets and its place within these change, its manufacturing structure and operations can become misaligned to its business needs. A change programme includes the realignment of a company's resources. The longer term problems are generally separated from quick hit opportunities.

The change programme is often a rolling programme, typically for 1–2 years, with clear targets for each period of operation and specific indicators to measure its success. For instance, Lucas Industries Ltd, a major UK manufacturing company, has a competitiveness achievement plan (CAP) which prescribes the priorities for the full time multi-disciplinary task-forces charged to carry out the defined projects. One dimension of achievement is to attain technological progress without the cost levels of 'high technology solutions'.

The task force is marshalled specifically for a defined project and is disbanded when the project has been completed. The right key leader is essential and good team dynamics need to be fostered.

Where much of the change programme has been carried out by staff departments, such as production planning or design, particular care is needed to involve the implementing departments of the company. These are the departments that have to work with the changed processes. If the staff of those departments have not participated in the change preparations, if they are used to 'steady-state' conditions, then, as far as they are concerned, the change process is external to them and is unlikely to obtain more than passive support. This is the first step towards alienation

13.10.2 Planning a plant improvement programme

In the process industries and increasingly with engineering production, manufacturing processes are based on a high degree of automation. While the utilization of operator know-how will always be important, further improvements depend increasingly on key process variables. Such indicators as throughput rates, energy efficiency or plant yields will suggest an improvement route. Yet with many process units the interaction of variables and second-order effects can nullify the expected improvement, indeed be counter productive. In a multi-parameter operation a portfolio of plant interactions has to be considered.

The first task with any endeavour to 'optimize' plant performance is to decide which parameter is the most important. This is not a once-for-all decision, because it depends on the economic value of plant inputs and outputs. For instance, process yields may be the most important objective, but when there is a fuel crisis and fuel prices rocket, energy efficiency may become the critical condition. The second set of decisions are concerned with the trade-off of competing parameters.

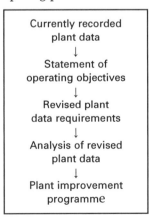

Figure 13.1 The planning task for a plant improvement programme.

In turn, meaningful decisions require considerable operating experience expressed by an appropriate reporting system. With computer-controlled systems the information is readily to hand, but, of course, will only report to instructions within the original systems design. The control systems will adjust and optimize plant operation to pre-set instructions. We are concerned here with the further improvement of plant operation, not envisaged by the original systems design. Figure 13.1 illustrates the planning task for a plant improvement programme.

13.11 TECHNIQUES FOR IMPROVEMENT

We have already noted the importance of free discussions, the relationship of trust and the welding of groups of people into effective teams. One yield of such a setting is greater stimulus, the generator of ideas; the creative processis encouraged. This is the point of departure for improvement programmes. Techniques can be divided into behavioural and analytical tools. There are several well known techniques that can be used to stimulate perception and to foster the novel association of ideas. The most relevant are given below.

13.11.1 Creative processes

According to Twiss (1986) the creative process is characterized by:

1. new combinations or patterns of existing or new knowledge and combinations arising from the imagination;
2. the association of ideas, often from widely different fields of learning, which allows new patterns to arise;
3. the redefinition of problems;
4. the stimulation of people with different intellectual backgrounds;
5. the putting aside of normal rational thought;
6. the questioning of assumptions;
7. the avoidance of premature critical judgement.

(a) Brainstorming

This technique has had considerable application. A mix of staff is chosen to discuss a specified problem or opportunity on a 'free-for-all' basis. The basis for staff selection is diversity of experience or function and the ability to make a contribution. The purpose of the discussion is to get ideas through the stimuli of discussion and personal interactions. The more ideas the better. A skilled chairman is important to gather the ideas and to ensure that participants freely give their thoughts. It is crucial that no judgement is passed at this stage on any idea that is forthcoming. After a period when no more ideas are forthcoming the collection phase is closed. The evaluation stage then follows where the obvious non-runners are put aside. The surviving ideas are then considered in more detail and collated in a portfolio for further assessment.

An alternative approach with ongoing manufacturing activities is 'negative brainstorming' where participants in a group discussion concentrate on product and process faults, fault possibilities and weaknesses. The resultant ideas are then sifted and assessed in the same manner.

(b) Value analysis

Value analysis is the evaluation of an existing product, process or system to see whether it can be improved. Essentially, value analysis is a critical but constructive and creative review where a functional specification is given, but the agreed function might be more efficiently and economically achieved in other ways. The study is concerned both with the overall function and the functions of each element or component. The review is periodic and one purpose is to incorporate technological advances into a product or process which were not available when the original design was developed. Typical benefits of value analysis are component simplification and standardization, the use of new materials which yield cost savings and the replacement of a group of components by an integral unit.

The following are the main constituents of the technique:

1. The information phase. This considers the functional specifications and their implications and such related aspects as design requirements, contract conditions and other constraints. It also ascertains the cost structure of the constituent parts.
2. The creative element. As part of information gathering, ideas will be generated and these are listed. In addition to this, brainstorming and other creative techniques are used to develop the portfolio of promising ideas.
3. The judgement phase. The listed ideas are then evaluated for advantages and disadvantages, after which they are rated for promise. Only those ideas which achieve a minimum rating will be retained.
4. The development phase. The surviving ideas will be further developed and their feasibility confirmed. The most promising are then costed in detail and their benefits quantified.
5. The recommendation phase. The findings are then submitted for management approval. Reasons for the changes and expectations of benefit are given.

A value analysis is a systematic task and companies often design forms to structure their approach and to evaluate choices. It is best carried out by an interdisciplinary group so that assumptions can be checked quickly and the stimuli of discussions can be maximized. Value analysis is a typical management technique. It yields the benefits of cost reduction and added value, but there are the costs of the analysis. The technique needs to be managed. Zimmerman and Hart (1982) give a good account of the technique and its applications.

(c) Evolutionary operation

Various statistical techniques are available for a plant improvement programme where the behaviour of a number of parameters has to be considered. A particularly useful technique is evolutionary operation (EVOP), which was first applied to the improvement of agricultural crop yields. Since then it has had considerable application in the process industries.

EVOP requires that slight, but systematic variations are introduced into process operating conditions. The variations are small so as to maintain good quality output. The introduced variations follow that of a conventional factorial experiment, so that, for example, with an investigation of two variables a design of four parameter value combinations will be run, such as:

high/high; high/low; low/high; low/low.

One complete sequence of the combinations is a cycle. As the variations are small it will need a number of cyles to eliminate random background variations. If significant changes are detected, they can be related to some of the objective functions of plant improvement. Once better settings of the observed variables are discovered the next phase consists of a batch of cycles around the upgraded settings until a limit is reached.

The technique can be applied in a systematic form with standard data recording formats which permit ready use over subsequent shifts. Considerable manufacturing gains can be obtained. In a case study from the cement industry, reported by Carter and Lowe (1974) a much fuller understanding by operators and technical staff of the actual process conditions was a useful by-product. What was previously a 'black art' became explicit and could be documented for training and plant operations.

13.12 THE ECONOMICS OF IMPROVEMENT

The drive for improvement is to achieve, or at least to maintain, competitive advantage, particularly in the form of cost reduction. In some industries, where component suppliers are under continuous pressure from large corporate customers, it has become a matter for survival. With global resourcing by such customers, a company without continuous cost reduction would lose its margins or its business.

The use of cost reduction programmes as improvement drivers is well known and some electronic companies are sufficiently confident to base their forward component price curves on these. For instance, one major international computer manufacturer aims at 5% per annum cost reduction in real terms. Such a target is based on experience and learning curves. It has been established empirically, particularly withibatch manufacture, that each time a firm's cumulative output is doubled, the total unit cost of such a product, in real terms, is reduced by a relatively constant percentage. The percentage is some industries has been as high as 20–30%. Apart from any economy of scale, the main contributories to such cost reduction are learning effects and the continuous drive for improvement.

13.13 SUMMARY

The mansagement of improvement is concerned with the more effective use of an existing set of resources and technologies. It stresses continuous learning and the constant quest for progress. Its basis is the culture of an enterprise, which emphasizes involvement, consistency of values, adaptability and a sense of mission.

Improvement is encapsulated in the Japanese term 'Kaizen' – a philosophy of life, applied to industrial operations, while the mission of 'poka-yoke' is the complete eradication of all manufacturing defects. Techniques, such as brainstorming, value analysis and evolutionary operation can be applied to manufacturing and to other functional activities.

FURTHER READING

Cyert, R.M. and March, J.G. (1963) *A Behavioural Theory of the Firm*, Prentice-Hall, New York.

Denison, D.R. (1990) *Corporate Culture and Organizational Effectiveness*, John Wiley & Sons, New York.

Imai, M. (1986) *Kaizen: The Key to Japan's Competitive Success*, McGraw-Hill Publishing Company, New york.

Kodama, F. (1991) *Analyzing Japanese High Technologies*, Pinter Publishers, London.

Nikkan Kogyo Shimbun Ltd (ed.) (1988) *Poka-yoke. Improving Product Quality by Preventing Defects*, Productivity Press, Cambridge, MA.

Tranfield, D. and Smith, S. (1990) *Managing Change*, IFS Publications, Bedford.

CHAPTER 14

Technology and standards

14.1 INTRODUCTION

The economic utilization of new technologies requires a structure and a pattern of usage to extract the expected benefits. It is like the discovery of a new seam in mining – the challenge is how to get the best out of it. The analogy is incomplete, because the benefits and costs have to be assessed at four different levels:

1. the level of the company,
2. the industrial sectors concerned,
3. the national economies affected,
4. the international context.

The different levels overlap and the trade-off criteria vary. Different interest groups are involved. For instance, the benefits a corporation wishes to secure do not necessarily lead to national advantage. There is also the dilemma of structures. They provide a public good but in a changing world their benefits can be eroded and their life cycle ends when they become an obstacle to further developments and are superseded by new structures. Standards and standardization form such a structure.

This chapter is concerned with the interactions between technology and standards. Technology permeates standards, while standards order the use of technology. Technological change affects existing standards and emerging technologies require new standards. They can shift the patterns of power between the various interest groups. Some standards consequently acquire political dimensions.

One aspect, stressed by Reddy (1987), is the globalization of the industrial market place which is the arena for competing world class manufacturing companies. International standards influence international trade, particularly where some of the rapidly growing developing nations adopt them as the basis of their national standards.

14.2 BASIC CONCEPTS

The definition of a standard as given by British Standard 0 Part 1: 1981 (quoting ISO Guide 2: 1980) is as follows:

> A *standard* is a technical specification or other document available to the public, drawn up with the cooperation and consensus or general approval of all interests affected by it, based on the consolidated results of science, technology and experience, aimed at the promotion of optimum community benefits and approved by a body recognised on the national, regional or international level.

The same document specifies standardization to be:

> An activity giving solutions for repetitive application to problems essentially in the spheres of science, technology and economics, aimed at the achievement of the optimum degree of order in a given context. Generally, the activity consists of the processes of formulating, issuing and implementing standards.

There is a wealth of implications behind these two concepts which can be approached in two ways. The first approach is 'normative' – standards are to be developed for the private and public good. The second is 'positive' – standards are what they are and reflect the balance of power between different interest groups. Both approaches are relevant here.

14.2.1 The functions of standards

Standards have a wide range of general economic and social functions of which the following are the more important:

1. A compatibility function. This ensures that one product is compatible with another; they can be used together. It applies to artefacts; a fire-hose coupling must be compatible with a fire hydrant. It creates interchangeability between products and components. It is also concerned with systems interfaces; computer software must be compatible with operating systems; interface standards are an essential condition for telecommunication systems.
2. An information function. This comprises terminology, measurement and test methods. It is concerned with the structured and efficient transfer of information.
3. A quality function. This establishes product and service performance criteria in terms of reliability, efficiency, durability etc.
4. A health and safety function. This provides the benchmarks for legislation concerned with health, safety and the environment.
5. A variety reduction function. This facilitates the rationalization of production and inventories by limiting type variety.

One major effect of standards is the reduction of risk. They provide structure and give assurance to the potential users of the products with which they are associated.

14.3 STANDARDS AND THE INDUSTRIAL COMPANY

14.3.1 The operational levels of standards

Standards can play different roles within a company and judgements .about their benefits need to indicate at what operational level they are considered. Within a firm they can be evaluated at the following four levels, as shown in Fig. 14.1:

- *Level 1*: Standards are problem-driven, e.g. by an oversized, disordered inventory which presses management to adopt variety reduction. They are also an *ad hoc* response mechanism to a technical procedure problem.
- *Level 2*: Standards meet a product/production requirement. Typically, they are islands of standardization in production or quality assurance departments.
- *Level 3*: Integrated systems of standards traverse traditional functional boundaries. Many are computer-based and link, for example, design, procurement and quality assurance.
- *Level 4*: Standards feature within every aspect of a business but particularly at a strategic level. Standards are considered as a marketing and export opportunity and feature as a component of world class competition.

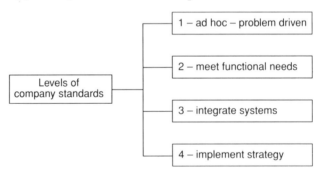

Figure 14.1 The operational levels of company standards.

14.3.2 Industrial standards and commercial transactions

The effectiveness of any arrangement between a purchaser and a supplier is highly dependent on the clarity of the understanding between them as to what is required. The outstanding function of a system of standards is to provide a language for technical communication which is adequately precise and comprehensive. Using this language, it is possible to define the properties of materials, processes for treating them and performance characteristics of finished products in terms which are widely understood.

The most obvious benefit to both purchaser and supplier is the enormous saving in time and effort when setting up a contractual agreement. But equally important is the consistency and reliability with which any part of a technical specification can be reproduced. This is a factor which is paramount in the control of quality.

14.3.3 Industrial standards and quality assurance

Standards and standardization have a close link with the quality of products and services. One reason for this is the success of the campaign to encourage firms to attain quality assurance certification under ISO 9000 or BS 5750. To satisfy the required assessment a firm must have written procedures for almost everything it does and it is often the simplest and wisest thing to base these on published standards wherever possible. For example, BS 5781:Part 1:1994, 'Metrological confirmation systems for measuring equipment' would be the basis for the company's calibration system. Similarly, a company manufacturing domestic smoke detectors would write its test procedures, using BS 5446, 'Smoke detectors for residential premises'.

14.4 TECHNOLOGICAL CHANGE AND STANDARDS

The notion of a standard is that of a benchmark which is fixed and stable. Other things which are uncertain, changing and variable can be tested against it. They will either pass or fail that test. A standard, therefore, provides the basis for a clear understanding between supplier and purchaser, between manufacturer and user. It is the aim of the standard maker to achieve precision and clarity. Success in this task promotes efficiency and agreement. Failure forebodes conflict and waste. It would thus be ideal if a well-formulated, proven standard could also be unchanging and enduring.

However important this criterion may be, it is even more important that a standard be relevant and workable. Relevance requires that a standard should represent the appropriate stage of technological development. A standard which specifies practices, considered by the interested parties to be out of date, will fall into disuse. Whilst there is a call for standards which are stable and enduring, there must always be the overriding proviso of relevance to the current state-of-the-art.

Standards can influence both attitudes and the rate of technological progress. The adoption (by consent) of exacting standards is reflected in the outlook of producers and in the expectations of users. It is also a spur to firms who do not reach the standard to improve the functions and the quality of their products. Similarly, standards which draw upon newly established technologies have a pace-setting capability which encourages innovation and the diffusion of the new technologies. For instance, standards for the measurement of, and the testing for, corrosion have encouraged the development of new anti-corrosion materials and coatings. However, standards which are based on tentative, unconsolidated experience may fail because they are premature.

The exploitation of new knowledge and an emerging technology often requires the updating of existing standards in the fields concerned. It can also cause deficiencies in standards coverage and additional standards will be needed. One purpose of such standards is to unify the efforts of many designers and producers, avoiding a proliferation of competing company standards. Lack of agreed standards can be disastrous for both user and producer. The user finds that articles from different suppliers are incompatible. A manufacturer can get out of step and be faced with the costly task of product redesign and the obsolescence of current output.

Where a major technological change occurs a new material, product or process replaces those currently in place and often yields a radical transformation of practice. The standards which express the existing practice very quickly become obsolete.

No technological change is ever final. It will simply begin a process of evolution as manufacturers apply their own skills to its exploitation. Where the change consists of gradual, evolutionary developments then refinements to products, techniques and equipment take place continuously. However, there comes a stage when cumulative developments make the product or process significantly different from the original. The product or process is gradually 'outgrowing' the standards by which it was originally specified. Since this takes place during the diffusion of a technology in industry, such changes will generate a host of variants on the market unless standard makers promptly follow up this evolution. The dilemma for the standard makers, be they a trade association or a national body, is that they have to aim at moving targets.

14.4.1 Technologically related changes of standards

Apart from technological changes there can also be changes in the application of an existing technology. An example is the requirement that a petrol engine should run on either leaded or unleaded fuel interchangeably. This requires the alteration of engine test standards without calling for a new technology.

Another source of change emanates from the macro-effects of new technologies, such as the development of generic technologies. The use of computers and sensors has a broad impact on testing, measurement and calibration. Also emerging technologies with new hazards generate new health and safety legislation which calls for, affects or makes use of standards.

14.4.2 Standards and competitive technology

We have already seen in Chapter 2 that in the early stages of a new technology there is a high rate of innovation and product/process development. Alternative design concepts compete until a dominant design emerges. That design will be the strong candidate for a standard. If it is adopted its owner(s) will have an advantage as competitors will have design, manufacturing and marketing adjustment costs. Conflict, rather than consensus, will be the background of standard making until the choice is made. Of course, this is a specific scenario; there are other contexts where a relatively mature technology or a costly emerging technology encourages cooperation rather than competition.

Farrell and Saloner (1987) also indicate that existing standards for an established technology lead to inertia, especially where an industry is 'tied together' by compatibility. A switch into a new technology carries the risk of not being followed.

14.5 THE MAKERS OF STANDARDS

Standards are developed and managed by different organizations at various levels within an economy. A consideration of the activities of the various standards making bodies indicates the extent and complexity of the standards making and updating processes. The basic requirement of a standard is that it is acceptable to the user. That does not only mean that it shows technological competence but that there is a consensus about its contents. To achieve this can be a long process.

The following are the most important groups of standards makers.

14.5.1 The company

A firm of substance or one operating in a field of advanced technology often develops company standards. These may range from engineering or construction standards for its operations to product standards as part of a commercial strategy. Basically, standards yield benefits with large-scale operations having a high repetition factor. Mass production is the prime example. However, a company standards department is not just concerned with the formulation of in-house standards – that would be necessary when no suitable external standards are available. It is equally involved in adopting and adapting external standards and advising on contractual aspects where standards are called for by customers. An important but less publicized activity is the participation of its staff in national and international standards making activities where they can make expert contributions as well as promote the interest of their company.

14.5.2 The trade association

The aerospace industry, with its rigorous and specialist design and safety requirements, illustrates the role of a trade association in the development and management of standards. In Europe, AECMA (Association Européenne des Constructeurs de Materiel Aerospatial) is regional and in North America, AIA (Aerospace Industries Association of America, Inc) is essentially national.

AECMA produces aerospace standards for the European Union, which after ratification also become British Standards. Its members are the major West-European aerospace companies which are represented directly or through national trade associations. AIA, through its Technical and Operation Council, controls a number of committees, one of which is for national aerospace standards. This committee oversees and coordinates the writing of standards which are subsequently published by the US National Aerospace Standards Committee (NASC)

14.5.3 Independent standards bodies

Some institutions have evolved in response to commercial needs and technical requirements. A prime example is the American Society for Testing and Materials (ASTM). This is a voluntary testing and standards development organization which works on a commercial basis and without subsidy from the US Federal Government. The writing of standards is the responsibility of about 130 technical committees, which prepare standard test methods, specifications, practices, guides, classifications and terminology for materials, products, systems and services. ASTM publishes about 8000 standards annually.

ASTM regularly runs series of 'state-of-the-art' seminars where experts are invited to give an account of the latest developments in specialist fields. A key feature of ASTM's strategy is its Institute for Standards Research (ISR) The purpose of this institute is to improve the quality and timeliness of ASTM standards through the identification, development and management of research activities.

14.5.4 National standards bodies

These are the bodies which are either formally charged with, or have attained, the responsibility for standards development and management in a particular country. They are mostly independent agencies, but work closely with their national governments from whom they obtain financial support at varying levels. Traditionally, a substantial portion of their income was obtained from the sales of standards. However, in some fields of advanced technology, where only a few international companies would use key standards in their products or systems, there is now a trend for such firms to support the national body directly. In a sense, that body becomes an extension of a company standards department.

It is interesting that there is no national US standards making body as such. The National Institute of Standards and Technology (NIST), is a non-regulatory agency within the US Department of Commerce. Apart from standards for computer systems and mandatory standards, where it participates as a federal agency, NIST is not directly concerned with the making and updating of standards. Generally, the publication and management of standards in the USA is in the private sector; there are no less than 400 voluntary standards making bodies. Nevertheless, NIST has a very strong influence on standards making and its staff participates in about 500 standards committees. Its involvement with emerging technologies links it closely to the development of new product standards.

The following are the three most important European standards making bodies:

(a) The British Standards Institution (BSI)

The British Standards Institution is the UK national standards organization which operates under Royal Charter. It is a substantial organization, which in 1993 had about 2000 employees and a budget in the region of £100M. It is also a major certification body. BSI standards are prepared by over 1000 active, technical committees whose members are volunteers and seek to represent all possible interests. Standards are thus prepared as consensus documents and result from extensive consultation. The secretariat, composed of technically qualified people, is divided into subject areas and each division is responsible for standards in that subject area.

An important opportunity for contact is established through the composition of the six standards councils, their steering groups and the standard policy committees. Many members are involved in technical developments and their commercial implications in the areas concerned. These contacts are mostly informal, but they provide a useful communications network.

(b) Association Française de Normalisation (AFNOR)

AFNOR is an independent standards making body. Its standards are given formal status in France by government decree. Its standardization activities are divided into 19 sector programmes, each representing a major area of technology. A sector programme is the responsibility of a strategy committee, with members from government departments, trade associations and industry. At the steering level there are 27 Bureaux de Normalisation which work with specific industries. The actual standards making work is carried out by about 1200 technical committees. AFNOR publish about 1000–1200 new or revised standards a year.

(c) Deutsches Institut für Normung (DIN)

DIN is the most important standards making body for the Federal Republic of Germany. It publishes about 1300–1400 new or revised standards each year. Although of the highest international reputation, DIN faces significant competition within Germany from the Verein Deutscher Ingeneure (VDI), the German professional engineering institution which issues guidelines for the use of its members. These often become *de facto* standards.

14.5.5 Regional standards bodies

These are associations whose primary function is to integrate the standards for a group of politically and/or economically related countries. The most important in Europe are CEN, The European Committee for Standardization, and CENELEC, The European Committee for Electro-Technical Standardization. These bodies are concerned with the harmonization of national standards in such a way as to eliminate trade barriers and to rationalize legislation.

14.5.6 The International Organization for Standardization (ISO)

This is a worldwide federation of national standards bodies, which in 1990 had 87 members. The ISO objectives are to promote the development of standardization and related activities in the world, to facilitate the international exchange of goods and services and to develop cooperation in the sphere of intellectual, scientific, technological and economic activity. ISO is more concerned with technologies than products. It sees its role as a facilitator of technological developments where standardization, particularly of systems interfaces, supports the technologies concerned. Its authority to develop and publish international standards is based on the agreement of its member countries. More than 70% of the ISO member bodies are government institutions or organizations incorporated by public law.

The work of ISO is carried out by about 160 technical committees, each concerned with an area of technology. They are constituted by, and report to, the ISO council but are intended to determine their own work programmes and to carry them through. Coordination and collaboration with other standards making bodies and other organizations are encouraged. In order to speed the standardization process, the main committees divide their subject areas and assign specific parts to sub-committees. They also establish working groups to deal with different aspects of a technology. For example, a main committee, JTCI, coordinates the work of about 20 sub-committees over a wide area of the information technology field. More than 60 working groups are associated with the subcommittees of JTC1.

ISO is closely linked with The International Electrotechnical Commission (IEC) which concentrates on the development of international standards in the field of electro-technology and electronics.

14.6 THE EFFECTIVE MANAGEMENT OF STANDARDS

Although we have already seen that standards can be formulated and used at all levels of economic activity, the attention here is focused on the national, regional and international levels. This does not overlook the fact that a leading company in a fast-moving technology may develop a standard which is subsequently endorsed by national and international bodies. But whatever the precise circumstances, the effective management of standards requires the following

1. that the standards needed by prospective users, such as for the efficient formulation of contracts, are readily available to them;
2. that what is available is communicated adequately to the potential users;
3. that users can receive some support (e.g. on questions of interpretation) in using standards;
4. that the standards specify good current practice.

14.6.1 Problems of timing

Timeliness is always important with a new standard. A difficult case, requiring a high degree of judgement, is associated with technological change. The problem is at its worst when changes are evolutionary in character but progress is rapid. Premature standardization or updating will give rise to frequent amendments or revisions which is a source of confusion and error. Delayed standardization makes the process of alignment or re-alignment to the new standard very difficult.

Where international or regional standardization on a given subject is anticipated, national standardization bodies may prefer to adjust their work programme and wait for those standards.

14.6.2 Standards and investment in innovation

While not necessarily a primary factor in company investment decisions, standards can affect the nature of risk and the likely profitability of investment in new production equipment. Established standards with agreed test and verification procedures reduce technical risks and enhance the market acceptance of new product innovations. A report to the US National Bureau of Standards by Putnam, Hayes and Bartlett, Inc (1982) quotes the case of compatibility standards of automated equipment for the manufacture of semiconductor devices. No manufacturer made a complete line and the users of the equipment, the device manufacturers, had to be sure that the equipment from different manufacturers was compatible. In turn, the equipment suppliers could develop the automation equipment because wafer dimension standards had been established. These standards facilitated the automated alignment of wafers which substantially increased production yields and simplified the development of test equipment. Also wafer carrier standards were introduced as these were being developed. The resultant improvement in quality permitted the use of larger wafer sizes with corresponding major economies of scale. The standardization of key wafer dimensions and carriers reduced the risks and accelerated the diffusion of the new technology. It also reduced the technical risks of developing very large-scale integration (VLSI) which calls for extremely accurate measurement and test methods.

14.6.3 The extent of detail

Standards are essentially prescriptive; they tell the user what should be done. However, there is a limit to what can reasonably be prescribed. A standard has to make assumptions about the competence of its users and the accepted practice within an industry. There can, therefore, be a problem about the amount of detail required in an area of rapid technological development.

An interim standards will often allow considerable discretion to the user and encourage him to furnish information which forms the basis of a more detailed specification. Generally speaking, over-specification may limit the usefulness of a standard in applications which originally were not envisaged, whereas the lack of essential detail can impair the usefulness of a standard as a source of guidance.

14.6.4 The lack of a corporate standard

The lack of a corporate standard for communciations protocol within the world-wide General Motors organization resulted in some fragmented investments. Each manufacturing plant in Europe had purchased various manufacturing hardware, running different operating systems, controlling similar applications. The company lost two opportunities:

1. a single supplier hardware purchase agreement could have provided volume discounts as well as plant compatibility;

2. with common platforms any investment in application development would have yielded maximum return as it could be adopted at any General Motors site with minimal customization.

The experience was a factor behind the US development of a common manufacturing automation protocol (MAP) which sets the standard for software as opposed to a hardware platform level. Similarly, a set of standards have been developed for General Motors Europe for all application development on its plant information computer system.

14.7 THE ASSESSMENT OF TECHNOLOGICAL CHANGE

14.7.1 Standards related to the technology life cycle

It will be recalled from Chapter 2 that the life cycle of a technology can be divided into an initial fluid stage, a settled, mature stage and a period of transition between the two. In the fluid stage of the technology and its associated product life cycle the pace of innovations is such that product designs are relatively flexible. In this phase, standards can have a negative impact on the evolution of some product design and performance. It also may be difficult to formulate a standard within a competitive industry when market stakes are being established.

The development of standards and the timing of their introduction within the technology life cycle is a significant matter which is affected by the degree of consensus and competition between the interested parties. One noticeable development with advanced and generic technologies, where there is government supported pre-competitive collaboration, is the development of initial standards which facilitate development and application work.

14.7.2 Forecasting technological change

In the same way as a company technology strategy is based on the anticipation of technological developments, so a standards making body needs to have a view about technological changes to come. Naturally, it relies on the guidance of its technical committee members who are familiar with the state-of-the-art in a particular technology.

We have already noted in Chapter 5 the different possible approaches for forecasting what kind of developments are most likely, or for predicting how long it will be before a particular innovation triggers a change in practice.

Technological change which has made its mark on existing practices is relatively easy to provide for. It is far more difficult to anticipate a change and to pre-judge its effect on products or working practices and consequently on the standards relating to them. Although it may be unwise to produce standards ahead of a technological change, there is some value in being forwarned of trends. Progress can then be monitored and a standard developed at the appropriate time, not thwarting innovation and competition, yet limiting undue proliferation.

In a recent research investigation, in which the author participated, the reasons for new or revised standards were analysed, using a sample of 34 cases. Of the total, 18 were accounted for by rewrites of other standards. This is an interesting indicator of the interrelationships of standards. One was a revision/extension for a reliability and maintainability procedure. The remaining 15 had some element of embodied technological change and consisted of nine full new standards, one new interim standard and five revised standards. Some of the technological changes were concerned with the test specifications for rubber, new instrumentation for fuel tests, meeting new effluent requirements, storage specifications for a new powder coating material and the specification of shaft-angle to digital value for rotary encoders.

Some of these changes represented the aggregation of evolutionary developments. The encoder specification typified the case of a generic technology with wide applications and numerous variants. The speed and resolution capabilities of sensors had rapidly increased and their reli ability improved. Multi-channel applications had become common.

14.7.3 R & D and standardization

There is often an awareness gap between research managers and standards makers, yet much of product and process development impinges on the tasks which form part of standardization. There is scope for cooperation between these functions. For example, work on test methods, the categorization of materials, codes of practice, performance criteria and design guidelines are common to both. An agreed terminology has to evolve. There are long-term benefits in including standardization as part of a development work programme. These tasks can be carried out in a standard way (such as recommended by BS 7000: Part 1 1989 'Guide to managing product design') which accelerates development work itself and application engineering. The process of diffusion of a new technology is considerably affected by the standardization process.

An interesting illustration of this is the field of advanced technical ceramics related particularly to high performance applications in engines and gas turbines. Initial work by the Japanese Industrial Standards Committee (JISC) and BSI concentrated on the standardization of current methods. A contemporary programme by Committee C–28 of ASTM was concerned with the performance, properties, processing, design, evaluation and characterization of advanced ceramics. A fundamental change in materials technology now made it possible to design materials to specification. This required a new classification, terminology, sampling and test methods to establish the physical, chemical, mechanical and thermal properties for ceramic powders, monolithic and composite ceramics and ceramic coatings. In essence, standardization was concerned with the survey and the taxonomy of new materials and the tools that were needed for their proper analysis.

14.8 THE DRAFTING, PRODUCTION AND PUBLISHING OF STANDARDS

As an example of the work of a standards making body a brief account is given of the relevant activities of the British Standards Institution. The underlying principles of operation are similar for all standards making bodies; most of the differences are in the detail procedures.

The definition of subject matter is one of the first decisions for standards making bodies and these can have far reaching implications. The scope and progress of their work programmes are dependent on the staff resources made available by interested parties at technical committee level and by the logistic support that the standards making body can provide. The opportunity cost for the development of standards in one subject area is the lower priority for standards making in other areas and the consequences that flow from that. The allocation of scarce resources will involve trade-offs and compromises.

A common function of a standard is to specify a product, process or test. There can be several alternatives to choose from. Seldom will such choice be simple. The various contenders will have their strengths and weaknesses, so the decision will depend on the analysis of a whole complex of factors. The eventual decision can disadvantage certain interest groups and may only be reached after extensive negotiations. The path to consensus can be long and hard. From the announcement of new work, via an existing technical committee, to the publishing of a standard will usually take two to three years.

A technical committee is responsible for the promulgation of a new standard. Usually members of the committee represent the relevant manufacturers and actual/potential users of the products or systems for which the standard is to be developed. Members are also appointed from appropriate research organizations, professional bodies, trade associations and sometimes government departments. Decisions are usually based on consensus. The committee is serviced by a committee technical officer and such supporting staff as may be required.

New work to be started, reviews to be undertaken, draft documents available and proposals for deletion are all announced in *BSI News* with an open invitation to comment. When the complete process of drafting, consultation and consensus is completed the standard is issued. Where the state-of-the-art is not yet fully established and there is insufficient practical application a 'draft for development' may be published instead. Such a draft may also be issued when it is judged to be too early to introduce a definitive standard, but guidance is still needed or an urgent requirement precludes the whole process of consultation. A draft for development (DD) used for this 'short cut' process is often referred to as an 'intercept standard'. The DD is used in certain areas of technology to expedite the standards publishing process.

Furthermore, the boundaries of a prospective standard also have task implications and where the subject matter is extensive, a standards making body may have to publish parts of a standard by instalments over an extended period of time. There is the further dilemma of the level at which a standard should be pitched. This applies particularly to design and test requirements. There is the choice of proposing a high standard for possible relaxation after comment or of seeking early consensus which tends to create a standard which is the lowest common denominator. The issue for the standards making body is how to assess and to secure the maximum overall benefits from standardization activities in the face of diverse commercial and public pressures.

Although the drafting task may ostensibly be simpler for tentative specifications or interim standards, the same considerations will nevertheless apply. A new technology, new product or newly discovered hazard can make a strong case for the development of appropriate standards. However, considerable research and experimentation may be needed before suitable test methods can be established or the credibility of a standard confirmed. In some fields there is an insufficient body of knowledge for a standard to become authoritative. It will therefore be appreciated that decisions about where to apply standardization efforts have to be based on a range of commercial, technical, safety and other factors and must be made within the constraint of available resources. The decision makers also need to consider the activities of other standards making bodies. To avoid duplication of effort it can be worth adopting a standard prepared by other organizations.

Furthermore, there are often strategic and commercial reasons for the development of a standard. A national standard which, in substance, is adopted by a regional or international standardization body has wider currency and brings with it marketing and certification opportunities. The reverse of such a situation and conformance to foreign standards can involve extra design and manufacturing costs.

14.8.1 The accelerated introduction of standards

In many ways the making of a standard is similar to the manufacture of a new product. There is a lead time for delivery, which is a function both of capacity and the rate of working. Capacity, in turn, depends on the level of available resources, which have to be scheduled to some priority rules.

BSI use a computer-based project management information system (PMIS) for their task scheduling. On occasion, they have made use of external secretariats and also contracted out the first draft of some prospective standards when there is a heavy demand for new/updated standards.

For some areas of rapid technological change, such as information technology (IT), there is also the 'fast track' mechanism operated by the International Standards Organization (ISO). An ISO standard is adopted and published within a BSI format under a single BSI/ISO number.

Major trade associations, such as AECMA, also have procedures for rapid standardization. The full normal process of formulation and international agreement amongst member organizations of AECMA is rigorous and time-consuming. Standards which have been published and are waiting for adoption by CEN are designated pr-EN and have the status of draft European standards (EN), although further comment or discussion is not intended. There is also, however, a 'Green-Paper' procedure by which a standard can be drafted quickly for consideration by the appropriate technical committee. The draft is distributed and committee members' comments are appraised. With such comment taken into account, the green paper is published, not necessarily conforming to all the editorial presentation practices. Its use is reviewed after six months, when it may be transferred to the full standardizing process or it may be modified, perhaps totally rewritten, and re-issued as a Green Paper.

14.9 THE UPDATING OF STANDARDS

14.9.1 Formal procedures for updating existing standards

Reviews of British Standards normally take place after five years. Drafts for development are reviewed after two years. It is, of course, possible for a standard to be reviewed at any time, should a sudden development warrant this. Some standards in fast changing fields are reviewed regularly at shorter intervals. The request for an early review would mostly come from a member of a technical committee – who would, if necessary, be invited to draft a proposed revision. Between regular reviews another procedure both aids the consultation process and reduces delays. A user who finds that a standard does not deal with his particular need is invited to put his problem to the technical committee who would give an answer or interpretation. The questions and answers are then published for the benefit of other users and will be taken into account at the next review.

In many cases the revision of a standard follows the same procedure as the original drafting. A straightforward review of a standard might be accomplished in one year but two years is more likely. When a standard is examined by the appropriate technical committee, the extent of the required change is the major determinant of the updating process. If the required change consists of only a few and relatively minor amendments then only these need to be issued for public comment. When more than five amendments have been made, a revised edition of the standard is normally published. Where the changes are of substance then a full review of the standard will be required and, although the work content of the review might be less, the same procedure as for the drafting of an entirely new standard will apply.

Where a complete review of a standard is needed, the staff resources required will become important and the priority of the work will have to be assessed within a portfolio of other standards making tasks. It is the responsibility of the standards policy committees to establish a balanced forward work programme and within this framework specific technical committees will determine actual target dates and plan work schedules. The overall speed of response of updating a standard is therefore a function of resource allocation as well as the time span of the standard making activity.

(a) The stimuli of technological change

At the technical committee level, the secretary of a group of committees keeps a running file of all the comments made about a given standard. He is expected to use his judgement in passing these to the appropriate technical committee and thus can be a focus for the stimuli of technical change. The secretary is technically qualified and his role is seen as that of a project manager. The members of the committee which he services are representatives of manufacturers, product users, trade associations and research establishments, who provide direct channels for the flow of stimuli.

14.9.2 The frequency of revision

The simplest system to operate is to have a universal, fixed period for the routine review of standards. This ensures that no standard goes unchallenged for longer than that period. A review, of course, does not necessarily mean that a standard has to be updated. But there must be a provision for a review at any time in the light of user experience and comment. As we have already mentioned, because of their more tentative nature, interim standards are normally reviewed within a shorter time-scale.

An important problem in some fields, such as electronics, is that the time needed for the normal procedures of revision exceeds that for significant technological change, i.e. the review process is no longer fast enough. Consultation, the basis of acceptable standards, is an inherently slow business. Speedy updating implies either accelerating the consultation or curtailing it.

14.9.3 The cancellation of standards

The review of a standard can reveal that there is no longer a need for it. A standard can become redundant as a result of a policy decision or a technological change. As a technology becomes obsolescent, so do the standards concerned with it. An important aspect of the updating process is the cancellation of such standards. This needs a certain amount of judgement because a superseded technology may still be used by some firms and spares could still be required for equipment in place. The withdrawal of a standard makes these a special supply, whereas its redesignation as 'obsolescent' allows it to remain on the publication list.

14.10 SOME SPECIFIC ASPECTS

14.10.1 An example of complexity – interface standards

These standards are concerned with the compatibility of computer-based systems, both within companies and world-wide networks. Where a system has no compatible interfaces it is isolated and in a commercial sense it will fail. For instance, with a computer-integrated manufacturing system (CIM) the special costs of tailoring interfaces between one set of production plant and another would be prohibitive. The prospect of such penalties has been one reason why there has been close liaison between standards makers and researchers at the research planning stage.

Again, electronic fund transfer (EFT) needs standard signals that carry information. There also has to be standard design of the operating equipment and standard software systems to ensure compatibility and effective connections. Particularly with telecommunications there is a strong case for standardization and simplification. Especially the smaller users, who do not have the power to affect the direction of the main service providers, have a need to work with standard protocols and interfaces so that the dangers of lock-in, wasted investment and expensive readjustment are minimized. User organizations often press governments and international regulatory agencies to enforce compliance with open standards, despite the difficulties of defining these quickly enough to keep up with technical developments.

(a) Open systems interconnection (OSI)

In the context of technical change, OSI is of special interest because developments have taken place very rapidly. In the past many users of computers encountered problems as new equipment or extensions to systems were incompatible with their existing equipment. Alternatively, software in use in a firm failed to run satisfactorily when new equipment was introduced. Mainly as a result of consumer pressures a complex, but effective, system for standardization has developed. It impinges on nearly all areas of information technology.

The organization and practices for OSI illustrate how collaboration and liaison is achieved internationally and how, in a fast moving field, standards can be produced and updated quickly. The main parties concerned are the manufacturers and users of computers, software houses and the network operators. ISO is the focus and the aim is to produce ISO standards for immediate national adoption.

The OSI programme started in 1977 and in 1984 the 'Basic Reference model for OSI' was published as ISO 7498. The model divides its subject compass into seven 'layers'. The ISO–IEC joint committee JTC1, Information Technology, has sub-committees assigned to each of the 'layers' of OSI with secretariats from the national standards bodies. In addition, there are many other sub-committees and working groups for highly specific topics. Thus, one contribution to rapid progress is the involvement of many small groups, each with a relatively narrow part of the total field and reporting to the appropriate layer committee. The remit of each group is to devise rules or 'protocols' for accepting the 'product' of its preceding layer, performing the required 'service' to it and presenting the result to the succeeding layer. Standards can thus be developed quickly.

A second feature of the OSI standardization is the use of the 'intercept strategy' as specified in the publication *Standards for Open Systems Intercommunication*, IT series no. 12 produced by CCTA, 1986.

(b) The use of intercept strategy

Although international standardization takes a long time, the main technical content of a standard sometimes becomes stable quite early in the life of a new technology. Where political acceptance is also widespread, a standard can – with some small risk but very considerable advantage – be introduced into service before completion of all the acceptance procedures.

For instance, the IT Standards Unit of the UK Department of Trade and Industry has formulated an 'intercept strategy' to identify and promote standards at the stage of near-agreement. The main criteria for making an intercept recommendation are:

1. widespread agreement on the function of, and the need for, the standard;
2. a high degree of technical agreement internationally;
3. a credible source;
4. and an assured future for the prospective standard.

Intercepts are recommendations, not standards; they are published as precursors to future standards anticipated from BSI or ISO.

Documents forming an intercept recommendation comprise:

1. the source document, usually based upon an ISO draft proposal or a document of equivalent status;
2. supplementary technical guidance, detailing options and modes of use.

Intercepts facilitate the quick introduction of protocols and recommended practice. However, other implications of an intercept strategy must be noted. Narrow specialization and many working groups impose a burden of coordination. Excessive use of the strategy results in too many draft documents which in any case are due for review within two years of publication.

14.10.2 The problem of manufacturing know-how
The surface mounting of electronic components

This case illustrates the need for a standard, where the knowledge to produce it is not publicly available. The surface mounting of electronic components on circuit boards has significant production cost advantages over the technique of through mounting. However, with surface mounting there is a problem of components coming loose, especially where a board is subject to vibrations or fluctuating temperatures. Manufacturers know that, even if they adhere strictly to the highest existing standards, their product is liable to premature failure. What is needed for the industry as a whole is both a method for making solder joints with adequate mechanical strength and a test specification which will ensure that such a joint has been made. Yet a particular manufacturer will develop a solution which yields competitive product advantage, which he may be reluctant to divulge. The solution techniques become proprietary know-how and an instrument of competition, possibly with patent protection.

At the macro-level this subject is of wide interest and a national or international standard is appropriate. But the standards makers cannot proceed until proven techniques are available. They rely on manufacturers or research establishments.

14.11 SUMMARY

Essentially, a standard is a technical specification which provides guidance on agreed practice and design. Its main functions are to ensure compatibility, to provide information, to establish quality and to limit type variety. For an industrial company, standards have a role at four levels: to resolve problems, to meet functional needs, to integrate systems and to implement strategy.

As a standard normally reflects a given state of technology, technological change impinges on its role as a benchmark. The corresponding updating or replacement of standards is a major task for standards making bodies. As standards affect the development and diffusion of new technologies their timeliness is important.

Standards are based on consensus and this may be difficult to obtain where there is strong competition within an industry. In those cases, the definition of standards is both a technical and, at times, a political activity when alternative designs compete for adoption.

FURTHER READING

Department of Trade and Industry (DTI) (1982) *Standards, Quality and International Competitiveness*, Cmnd 8621, HMSO, London.

Gabel,H.L. (ed) (1987) *Product Standardization and Competitive Strategy*, North Holland Publishing Co., Amsterdam.

International Organization for Standardization (ISO) (1982) *Benefits of Standardization*, ISO, Geneva.

Toth, R.B. (1984) *The Economics of Standardization*, Standards Engineering Society, Minneapolis

CHAPTER 15

Technology: an instrument of competition

15.1 INTRODUCTION

Chapters 4 and onwards have been concerned with the effective use of technology by an industrial company. We have seen that success here will enhance its general competitive position. However, there are aspects of competition where technology can make a particular contribution. The arena here are the global markets where product clusters are engaged in world-wide competition. Technology can be used to enter these markets and to establish a substantial share in them. For those already in this position the instruments of technology can be marshalled to defend such a position. This chapter is concerned with the role of technology both in securing and in maintaining competitive advantage. Some of the strategic aspects of entry and defence have already been covered in Chapter 4, but they are some operational and tactical aspects which also merit attention.

Already in the 1930s, Schumpeter (1934) argued that in a dynamic economy it was technological competition that really mattered. Technology provides one of several important constituents of the competitive advantage of a company. One fundamental opportunity which technology provides is the basis for added value and new products. This can be decisive in certain markets. Technological competitiveness is not based on cheap labour but on specific capabilities. With some advanced technologies it is the performance specification, not the price, which is crucial.

On the other hand, competitive technology on its own is insufficient. It needs the back-up of an integrated business strategy, corresponding business and professional skills, effective management, a continuous quest for improvement, leading to world class best practice, and at all times careful attention to the real needs of customers. Furthermore, it includes a readiness to work closely, and have a learning partnership, with suppliers. Technology competition has to be conducted in total systems terms.

As Coombs, Saviotti and Walsh (1992) have pointed out, there are several forms of technological competition. The first is within the same design configuration, such as between diesel engine manufacturers. The next is between design configurations, such as between diesel and petrol engines. The third is competition between different technological regimes, incorporating different design configurations and based on different sets of design principles, such as between a fuelled and an electric car. The different underlying knowledge bases make a shift from one to the other a major challenge.

15.2 SECURING COMPETITIVE ADVANTAGE

15.2.1 Product strategy

Where neither products nor markets are homogeneous, product strategy will be developed in terms of: product characteristics, market segments, price and product volume, marketing and technological development. For instance, Jacobsson (1986) notes that product growth can be characterized by the penetration of new markets through:

1. the standardization of the technology; fewer customer-designed features;
2. product differentiation, broadening the range available to customers, including simpler types;
3. price reduction, because of product simplification and economy of scale.

In his review of the international computer numerical control (CNC) industry, he gives some telling illustrations of different product strategies. For instance, a number of Japanese machine tool manufacturers embarked on an overall cost leadership strategy with the objective of penetrating the major sections of the engineering industry anywhere in the world. They achieved this by the product differentiation of their CNC lathes; their standardization and an increase in product volume to obtain scale economies. A key feature of their strategy was the design of lower performance, smaller and lower-cost machines, mainly for small- and medium-sized companies with limited technological staff resources. The Japanese machine builders focused on design simplification and design cost reduction, whereas European and US firms emphasized advanced technology and performance. The successful implementation of this strategy yielded a majority share of the world market in CNC lathes. The core technology was available at relatively low cost and suited the requirements of many users.

The life of a product is affected by the rate of advance of the technology on which it is based. When the technology is sophisticated it will cease to be so when it is superseded by subsequent advances. A core technology, incorporated in a volume product, can survive for longer where it has significant price and use advantages.

However, in those cases where there is a rapid change of technology and product lives are consequently shortened, getting the design and product 'right first time' becomes critical because there is only limited time for remedial changes and the re-establishment of customer goodwill.

Where companies operate with mature technologies, the lack of major technological breakthrough opportunities is pushing them towards incremental product improvements which are cheaper, easier and quicker to introduce. Of course, they are also easier for competitors to copy. To make this approach effective in a competitive context requires strong engineering and design support. Such resources in depth need to be developed as part of an integrated business strategy.

15.2.2 The product attribute spectrum

The function that a product performs is known as the 'product function'. It achieves this with a particular portfolio of product attributes which essentially are its features and characteristics. Those attributes determine the product's performance, i.e. the type and quality of the services it provides and their resultant value to its purchasers. Individual attributes are perceived to have different values and it is possible to rank these in customer value terms. This normally features in design specifications and analysis.

'Product differentiation' is an explicit portfolio of attributes which is distinct from those offered by competitors. It is possible to add functions to a given product. A simple example is that of a precision watch which, apart from its role as a timepiece, can acquire the function of a piece of jewellery. Technology can similarly extend product functions, such as the addition of a calculator to an electronic watch. Another form of differentiation is the change of the component architecture in a product, such as the dashboard rearrangement for a car.

'Product simplification' is a strategy for establishing cost leadership through a programme of standardization and greater modularity in product design. This reduces complexity and costs. With capital equipment, such as machine tools, this can also increase the value to the customer through easier operation and simpler maintenance.

'Product support' expresses the back-up which makes a product acceptable in use to a customer. This includes equipment service, instruction manuals, the ready availability of spares and updating facilities.

The market needs for a product function have normally a much longer time span than the product which currently serves that function. To meet these needs in the longer run requires a total product strategy and integrated product portfolios covering the main market segments. This can be assisted by 'product tack-ons', a key component strategy or packaging. For instance, Japanese electronics companies have been careful to retain the use of state-of-the-art components to provide a competitive edge for their products. This is crucial for short product life cycles.

15.2.3 Product attribute substitution

The product attributes must emphasize what the markets consider important. This means that the company has to understand how its customers actually think rather than how they are supposed to think. Technology is a contributor to such attributes. When market preferences change so must the development of the technologies involved. For example, after the 1973 oil crisis, the need for fuel economy for aircraft engines became much more important to airlines than higher aircraft speeds. This led to a re-direction of R&D programmes and the development of the 'lean' engine. Also there will be environmental changes, such as safety legislation and, with consumer durables, even fashion effects to consider.

15.2.4 Customer perceptions

The perception of quality and other product attributes, such as the capacity to solve specific user problems, is a complex issue. It is these attributes which determine the value of the product in the mind of the customer and indicate product added value. Although the criteria of valuation may be different, this applies to capital goods, raw materials, intermediate and consumer products. Furthermore, there are associated attributes, such as the effectiveness of service, technical back-up or the ready availability of spares.

Customer expectations can be affected by the marketing and technology strategies of companies. This has been very telling, for instance, in the personal computer business where technological advance and a stream of new products has engendered an expectation of continuous progress.

15.2.5 The fusion of process technology with distribution

The convergence of several technologies, i.e. in manufacture and materials handling, gives opportunities for competitive advantage. Process innovation is not confined to manufacturing operations. It includes everything up to the point-of-sale.

For instance, a leading chocolate confection manufacturer has automated its box and pallet handling to integrate production with automatic warehousing and bulk distribution. The packaging at unit box level was integrated with product promotion at the point-of-sale. Cost advantage and product differentiation were just as important within the distribution function as the manufacturing activity.

15.2.6 Advertising technological strength

While in the marketing of machinery and other capital goods it is common to refer to the strength, sophistication or robustness of the embodied technologies, this is less so with consumer goods where the appeal to customers stresses different product aspects. But even here a company with high performance products can use its technological reputation for marketing purposes. Translated broadly as 'ahead with technology' the German 'Vorsprung durch technik' advertisement, a key theme for Audi cars, is very explicit about the competitive role of its automobile technology and the extra satisfaction it promises prospective buyers.

15.2.7 Acceleration of product flow

Where competitors use shortened product life cycles as a marketing weapon there will be a pressing need to have new products in response. A company has the choice of escalating its product development commitments or deploy a strategy of product development flexibility with some potential products in the wings to respond to competitor strategies. This can be a costly and risky business, but could maintain a company in a particular market segment.

By concentrating on new product development, some Japanese companies have been able to maintain employment levels and profitability despite a 60% drop in their original business. Emerging technologies are rapidly incorporated into new products which dominate such markets as cameras, motor cycles and electronics. Product strategy concentrates on focused market segments.

15.3 TECHNOLOGICAL COMPETITION ANALYSIS (TCA)

The decisions whether to enter or exit from a market or how to develop a competitive strategy within an established field, require some very specific technological analysis. Glasser and Contino, with the Engineering Management Program of Drexel University, Philadelphia, USA, developed in the early 1980s this particular analysis which combines a form of technological forecasting with competitor assessment.

To start with, they distinguish between functional competition on products and processes from outside an industry and the competitors within the industry. Functional competition often results from technological advances outside the expertise and industrial know-how of a company. Its potential promise and impact is therefore more difficult to assess.

In most industries the main competitors know much about each other and their products. For a meaningful first comparison the analysis starts with a precise functional specification of the company's products based on field test data. The chief competitor products are then analysed and their functional performances established. At the same time the evaluation will yield many indicators of the manufacturing processes which have been used for component manufacture and product assembly. Patents are, of course, a significant contribution to research and development intelligence. The development of information technology, particularly databases on CD–ROMs, (compact disc read only memory) has facilitated this form of analysis. At the same time technical personnel can gain useful insights from carefully chosen visits to customers with company marketing staff. A systematic approach over a period of time will facilitate the building up of track records and enable the company to anticipate the moves of competitors.

The TCA technique is structured on nine stages of evaluation which are based on the following three matrices:

1. the current product matrix,
2. the futures matrix,
3. the result matrix.

In each case the firm is related in the analysis to its main competitors.

The current product matrix consists of the following:

1. The determination of product status. The objective of this stage is to compare the company with its competition in the technical development process. Global benchmarks are established to show the nature and the size of the gaps that exist. The benchmark analysis can be expressed most tellingly by a spider diagram such as Fig. 15.1 where each leg represents a technology attribute. The chosen attributes are those important to the company and will vary with different industries.

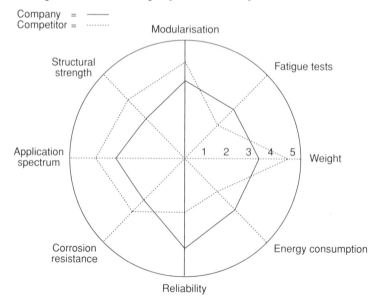

Figure 15.1 A spider diagram of product technology attributes.

2. The determination of 'near future' product/process capabilities. The time horizon for this is 2–5 years. This stage compares future performance expected by customers with the capabilities of competing products and processes. An example of this are the expected fuel consumption rates for future aircraft engines. It also involves an estimate of the likely time, costs and risks associated with achieving customer expectations.

3. Near-term fit with technological infrastructure. This includes the identification and categorization of driving and supporting technologies; also the time and costs involved in their development.

4. Present product/process ability to meet future market needs. This involves the decision whether to concentrate on the upgrading of current technologies or to adopt/develop entirely new ones.

The futures matrix contains the following three evaluation stages:

5. The assessment of the natural limits of current driving technologies and their hierarchy of criticality. The assessment has to establish which are the most critical driving technologies and how close they are to their natural limits.

6. Long-term technical predictions for the competing products or processes in terms of the driving technologies. This involves the study of each of the key technologies and the anticipation of their critical problems and natural limits. Their development paths are expressed in terms of rates of development, branching points, cash and time requirements. This is akin to the development of a technology road map.

7. Forecasts of future market needs and the ability to meet them. Company prospects are stated and compared with competitors. This includes predictions of the choices available to all suppliers. The forecasts incorporate the implications of environmental and legal constraints.

The final two evaluation stages refer to the result matrix.

8. The integration of technical projections with other market factors. This features non-performance customer expectations, such as price or style, to develop a market view. From this the following are determined:
   (a) cost performance discrepancies,
   (b) technological weaknesses,
   (c) the need for major technological steps,
   (d) the resource allocations required to achieve this,
   (e) the risks of doing nothing.

9. The definition of the need and timing for innovation. The technological and market projections are now incorporated in an overall business view, based on the costs and yields of the established innovation options. These include the choice of deferment and a risk analysis based on expected competitor behaviour.

15.4 TECHNOLOGICAL LEADERSHIP

This section expands on some of the strategic aspects described in Chapter 4. Where a firm has unique technological abilities this will yield a corresponding lead over its competitors. It will be at its most telling when a firm can supply a product or service which nobody else can. There are few runners-up, if nobody else can meet an advanced technology specification and there is little scope for price competition. The second-best loses. A market entry can be successfully achieved with a functional product superiority rather than a low price. This is usually achieved by an intensive technology product. The hallmarks of this are:

* a high proportion of new technology inputs is used in its production, typically R&D achievements;
* the use of a substantial component of professional scientific and engineering manpower in the main process functions.

Where a company has a well-developed DDMM system (design, development, manufacture and marketing) the manufacture of a new product is proven before related research papers are released. This furnishes a well-timed competitive advantage when the product is launched. Before competitors have a proper measure of the new product the company is well into the next stage of product development and cost reduction. Catching up will be difficult. In some fields of rapidly advancing technology, the lead company which combines marketing skills with its technological strength can develop and protect its technology so well that it leaves little room for competitors. Its position is enhanced as it establishes the first, and possibly the only, product and technology standards.

Another aspect of leadership is the concept of 'domain know-how' put forward by Rouse (1992). This reflects the firm's understanding of the domain, i.e. the key markets where its products are used. Technology-based companies sometimes underestimate the need for such know-how. Failures can occur because cited benefits are inconsistent with domain values and priorities.

An interesting example of technological leadership in sound reproduction is furnished by Ryan (1984) who quotes the Dolby strategy of 'well-protected openness'. A careful patent protection barrier is first established, after which the company furnishes technical clearing facilities which invite its customers – some of which could become competitors – to submit new product designs for constructive discussion.

We have already noted the nature and opportunities of the technology-cluster strategy which is based on the development and strength of generic core technologies. It is the broad base and commonality of these technologies which gives a company its opportunities and leadership in the market place. The hallmark of a generic technology is that it can be combined with other technologies to give application opportunities. In turn, this generates a wide range of products and in some industrial sectors, such as consumer electronic products, leads to product proliferation. The rapid development of new products and markets enables the firm to operate several businesses, linked by the generic technologies. The company can acquire an aura of competent technology where success in one field facilitates acceptance in the next.

15.5 THE ADOPTION OF A NEW TECHNOLOGY

Where a company, with mature technologies and products in relatively static markets, wishes to revive its fortunes it can be attracted to the opportunities of an emerging technology. If such a technology is still at the pre-competitive stage, time can be of the essence before competitors are able to establish themselves in a market and to put up entry barriers. There are three key stages for the company to follow if it wants to harness a new technology. As we have already seen previously there are many constituents to each stage.

1. Perception. This means the awareness of the new technology by the appropriate level of management and by the functional specialists concerned. With a radically different technology the organizational location of the 'new technology sensor' is not always predictable. A company's style of management needs to encourage perception.
2. Assessment. The implications, prospects, likely benefits and associated risks need to be ascertained. Risk reduction strategies are important.
3. Absorption. This is the stage when the new technology is harnessed to make an effective contribution to the operation of the business.

On the other hand, there can be a case for waiting. To be the first to adopt a new technology and to launch it on the market has its risks. Twiss and Goodridge (1989) list several arguments for caution when the following occurs:

1. The performance of the new technology is still lower than the current technology. New products incorporating the technology are still immature and costly.
2. There is slow progress at the bottom of the technology 'S' curve.
3. There are high initial costs where the learning curve benefits have not yet been realized.
4. The new technology is still unreliable.
5. There is past experience of new technologies which did not come up to their initial promise.
6. Potential customers of the new technology know little about it and at this stage may be risk averse.

On the last point, Jacobsson (1986) gives the example of numerically controlled machine tools. Large companies with substantial numbers of technical staff are usually better acquainted with new technologies and are more prepared to commit themselves to novel and sophisticated equipment. Smaller firms with less expertise are likely to go for more established technologies, often incorporated in more versatile and simpler equipment. In that sense they constitute a different market segment which may not be developed at the early stages of a new technology.

15.5.1 Decision factors for the adoption of a new technology

Time-pressure can tempt management to take short cuts but this can result in an ill-advised decision. Moenaert *et al.* (1990) indicate the following factors which a firm needs to consider before committing itself to a new technology:

1. Internal resources: human, technical: R&D equipment and infra-structure; funds, exposure: experience with the new technology; organizational slack;
2. External resources: external human, technical and financial resources the company can draw on;
3. Technology variables: stability, availability, threshold level (minimum investment in development) analysability (the ease at which you can get at its fundamental principles); variability, complexity (the number of scientific disciplines on which the new technology is based), controllability;
4. Strategic values: cost/benefits; the company's technological position within its industrial sector.

Burgelman and Maidique (1988) give a good account of the choice aspects of technological leadership and followership. Basically the decision hinges on the type of competitive advantage sought by the company within a given industrial setting. Typically, it needs to consider the following:

1. the technological opportunities to influence cost or product differentiation;
2. the uniqueness of the firm's technological skills;
3. the expected continuity or discontinuity of technical change;
4. the reversibility of the required investments;
5. the required technological support from suppliers or customers;
6. the momentum benefits of demand or cost reduction (e.g. product sales in period B are a function of sales in period A because of an established market position);
7. the degree of first mover advantage; if these are small, continuous innovation will be needed to maintain leadership;
8. the licensing opportunities of the new technology.

15.6 MARKETING A NEW TECHNOLOGY PRODUCT

15.6.1 The timing of entry

A new technology and a new market is an opportunity where time can be a competitive weapon. The first entrant in, or creator of that new market has the prospect of a temporary monopoly. If the product attributes and quality are right, price is not crucial and margins can be good. But such positions are transient; the greater the prizes, the quicker and stronger will be the competition. To exploit the benefits of being first the firm will have to concentrate on effective development to ensure that the product is free from the imperfections which hamper market acceptability. Furthermore, there will have to be close coupling with manufacturing and marketing to achieve significant occasions of entry, such as a major trade exhibition. As the market develops, there is continuous learning; the prizes have their risks and the prompt adjustment of plans and, at times, the cutting of losses will have to be faced.

The moment of entry is a signal to competition. Where these are powerful adversaries, particularly with strong marketing know-how then time is short and the product initiator needs already to be at the next stage of product development.

15.6.2 Waiting in the wings

This avoids the risks of being first in a new territory, but there are also the risks of being second. If the initiator is experienced, aggressive and powerful he will be hard to follow. Patent barriers will have to be circumvented and that could cost considerable development work. Know-how has to be established and unless the firm already owned this in a relevant field, it will take time and resources to establish. The strategy of waiting will only be meaningful where there is a long technology life cycle and the firm can mount intensive development efforts. It must also have a general competence in related technological fields, be capable of very fast market launch response times and, not least, have a high level of competitive intelligence.

Another opportunity is furnished where product or process changes permit the leap-frogging of the new technology. They yield the first mover advantages to the second-generation technology leader leaving the previous leader with the sunk costs of initial pioneer investments.

There is also the case where a firm already has the new technology but does not bring it to the market place. The technology then becomes a hedge against the tactics of competitors and an otherwise uncertain future. Protective R&D will generate the technologies which wait in the wings, to be marketed when the company is threatened by a rival. Existing, successful product lines are left undisturbed for as long as possible.

15.6.3 Entry strategies

As a result of technology scanning or a strategic review, some major companies become aware of the high growth prospects of a new technology. If they are not already in the emerging industry and a move in that direction is within their business plan, they can devote substantial funds to buy a promising performer in the new industry. If the acquired company is small and lacks the resources needed for the perceived opportunity, they can quickly apply their technological and marketing expertise.

But even big companies need to contain their risks by collaboration with others, such as:

- external agencies: universities, government research laboratories, trade research associations etc;
- suppliers and customers, particularly where there is significant application work with the new technology;
- competitors, where the risks of going alone are unacceptably high for all concerned. In any case, collaboration in one field does not preclude competition everywhere else.

15.6.4 Entry barriers

A firm can enter a new industry or it may already be in the industry but wishes to enter a new market segment or a new market abroad. The nature and extent of a barrier is a function of the industry. When the industry is homogenous and its product characteristics and prices are similar, such as in the car industry, the entry barrier is to the industry. Where the industry is heterogeneous, such as the machine tool industry, there will be subgroups of firms with different structural characteristics. The entry into the industry is, in essence, an entry into a group and the ease of entry will depend on the characteristics of that group. Also, the move from one to another group is akin to an entry into that group; e.g. a manufacturer of conventional grinding machines moves to numerically controlled grinding machines. The greater the disparity of the groups, such as the typical skills required or the market structure, the greater will be the risks of transfer.

The successful entry into a new or existing market is conditional on the available resources. These are primarily technological, logistic and financial. Particularly with technological resources, such as experienced design staff, there needs to be a minimum level before an entry will be effective.

An entry barrier may be likened to a wall but that can be a misleading analogy. Its magnitude can vary with the different stages of a technology life cycle. Stoneman (1983) makes the point that where a new technology generates a new industry, small firms have opportunities because scale economies are not yet important, markets are volatile and product loyalty has not yet been established. With a mature industry, scale, efficiency and brand loyalty make entry much more difficult.

Disregarding those barriers which have been established in international markets for political reasons, whether overt or disguised, the barriers are ultimately economic – there is the cost of entry. However, in those markets where advanced technology is embodied in products there are particular hurdles to overcome, such as:

- the need for a substantial infrastructure of company research and development;
- a heavy R&D/revenue cash ratio, particularly with a fast moving technology, where product life is limited;
- Substantial uncertainty with product and process technology.
- Shortage of technological know-how, design skills and operating experience. Core designs skills can be crucial.
- Lack of effective technical sales and service support. Where the repair and maintenance function is crucial to customer acceptance then the lack of effective distributors able to give such support will be an effective barrier.
- Lack of support from 'leading edge' suppliers.

The development of the European Community (EC) has illustrated how technology can be used to limit the free movement of goods which is the basis of the European Single Market. Each member state has its own health, safety and environmental legislation and if a product made within one country did not match the detailed requirements of another it could not be sold in that country.

In 1985 the EC Council of Ministers adopted a 'new approach to technical harmonization and standards' which set out the 'essential requirements' in general terms which had to be met before a product could be sold in any Community country. The normal method of satisfying this requirement is now to manufacture in conformity with a specified European standard.

15.7 THE RETENTION OF COMPETITIVE ADVANTAGE

15.7.1 Continuous product and process development

The long-term competitiveness of a technology-based company depends on the systematic exploitation of its product development capabilities. Harvey-Jones (1993) stressed this point when he emphasized the continuous need to add further product value. This can be achieved by providing extra service or solving customer problems. The recent development of fitting an integral power plug on consumer electrical goods illustrates this. It saves the customer a task and reduces hazards in the home.

Such product enhancement strategy is best implemented by a portfolio approach, which balances the magnitude and risks of the intended innovation with its potential returns. Likewise, the manufacturing process capability within the company is fostered by the continuous development of key equipment and the management of improvement.

A less obvious but important aspect relates to technological alliances with other companies. The development of the company technological capability can be a condition to stay within the 'patent pool'.

15.7.2 Ownership of patents

It is fundamental with patent applications that their claims should offer the maximum protection to the key ideas. The monopoly conferred by the patent and the protection against infringement is a barrier to competitors and in extreme cases a strong patent portfolio can become an entry barrier into the industry concerned. However, the work of Mansfield *et al.* (1981) has shown the degree to which it is possible in some areas of technology for competitors to design around a patent and that patents are not the only way to protect a company's knowledge base. The very fact that a patent specification has to be published and contains claims by the applicant for the patent gives information and opportunities to a perceptive competitor.

While a small number of patents in a given field may well be vulnerable to circumvention, there is the more recent tendency, particularly by Japanese companies, to concentrate very sharply on a defined target area and then cover all aspects with patents. For instance, in the field of genetics some Japanese companies have gone to great detail in patenting laboratory techniques and processes in addition to basic patents. Competitors may be all but locked out.

15.7.3 Safeguarding of know-how

Operation and experience develop specialist know-how. Much of this is person-specific. Of course, there is always the risk of key personnel being enticed to join a competitor or of industrial espionage. The firm has some protection in law against this but effective counter measures are often limited and expensive. Some key know-how can be kept secret, for instance, by the suppression of all information. It is not unusual for a process recipe to be locked in a bank safe deposit. But there is also the relatively routine learning-by-doing knowledge which a competitor has to acquire with time and money. Catching up costs.

15.7.4 Pricing policies

A range of techniques are, of course, available here. Competitors can be challenged by price reductions, discounts, rebates, free accessories, extended guarantees and other price reduction equivalents. A differential pricing structure in various markets for high technology products can undercut local firms by setting prices at which these cannot recover their product development costs.

15.8 SUMMARY

Technology has become an instrument of global competition. Firms compete within the same technology and between different technologies. Technology is a determinant of product attributes which influence customer perceptions and affect product value.

Technological competition analysis is a technique which combines benchmarking with technological forecasting. It enables a firm to assess the current and prospective strength of its technology in the market place.

The adoption of a new technology and its incorporation in new products is a key attribute of technological leadership. The timing of market entry is crucial, particularly where strong competitors wait in the wings. Being first is no guarantee of success. Competitive advantage is better maintained by the continuous addition of product value and the management of improvement within the company.

FURTHER READING

Burgelman, R.A. and Maidique, M.A. (1988) *Strategic Management of Technology and Innovation*, Irwin, Homewood, IL.

Coombs, R. Saviotti, P. and Walsh,V. (eds) (1992) *Technological Change and Company Strategies*, Academic Press, London.

Harvey-Jones, J. (1993) *Managing to Survive*, William Heinemann Ltd, London.

Rouse, W.B. (1992) *Strategies for Innovation*, John Wiley & Sons, New York.

Ryan, C.G. (1984) *The Marketing of Technology*, Peter Peregrinus Ltd, London.

Twiss, B. and Goodridge, M. (1989) *Managing Technology for Competitive Advantage*, Pitman Publishing, London.

PART THREE

Technology and Government

CHAPTER 16

National technology policy

16.1 INTRODUCTION

The discussion so far has been about technology, its impact on, and its opportunities for the individual company. The setting in which this takes place, and for that matter where it takes place, is a function of the overall economy in which the company operates. That setting is affected by government, its outlook, its political philosophy and its capability. Government here is no longer confined to national boundaries, but includes consortia of governments such as the European Union.

At this level there are basic and difficult questions. Should a government have policies on technology and if so, what should they be? Of course, there are some basic and relevant policies. One of the major functions of any government is to protect its citizens, so the citizen at work and in his private life needs to be shielded from the hazards of technology.

We have also seen that some companies have prospered because their development and skilled use of technology gave them competitive advantage. Others have failed. Should a government, like a 'super-company' make skilled use of technology? Although there are many differences of scale and structure, it is often a major objective, both at the level of the firm and the national economy, to stimulate the development of new technologies. Many of the techniques and approaches to help a company can also help a country. But what a government does, involves politics and politics, among other things, are concerned about value judgements. It is not the intention here to enter this arena but to look at the aims and policies of those governments who wish to improve the lot of their people through the use of technology.

Many governments are committed to the welfare and improvement of the living standards of their citizens. That means at least the maintenance and, hopefully, the growth of the economies of their countries. One major determinant of the strength of an economy are its industries, both manufacturing and service. Technology permeates these. A government, concerned with economic growth – or even just with economic survival – cannot ignore the economic aspects of technology. It needs to have technology policies. What these are often consists of a mixture of political and practical judgements but in the end they must yield competitive advantage.

Just as with an individual company, technology policies need to be integrated within a total framework of government policies. Mutual reinforcement or, at least, harmonization with other policies, such as economic, social, educational and environmental, is more likely to increase national well-being.

16.2 THE CONTEXT

There is the view that one way out of an economic recession and stagnating industrial markets is to stimulate technological innovation. Economic growth is fostered when a number of emerging technologies broaden the future industrial base. Some governments have a major role in this and for many years have been active in the development of industrial, science and technology policies. The encouragement of innovation is only one of a number of policies available to a government to achieve its goals. There is also an interdependence between goals, with each policy area having specific components as well as common elements. Consider, for instance, the interactions between an innovation policy and fiscal, economic, energy, environmental or employment policies.

Apart from the swings of the business cycle there are the longer term trends. The changed economic position of the United Kingdom provides a good illustration. A long period of relative industrial decline has been masked by comparative prosperity. Then a stage is reached when there is a wholesale collapse of key industries. In contemporary language, those industries have been 'hollowed out' by international competition. There is understandably a concern that the neglect of new technology could accelerate this decline. It is hoped to reverse this trend by appropriate emphasis on new technology in national economic development.

16.2.1 Some international benchmarks

The same approach as described for a company in Chapter 4 is relevant to the comparisons of national economies. The following highlights some important features of the current international position. For instance, Table 16.1 gives some *per capita* levels of civil R&D for some key economies. It is based on an analysis by Warren (1992) using 1990 data.

Of course, the different populations of the various countries makes the contrast of actual expenditures much greater. While input is no guarantee of comparative output it would be naive to disregard such levels of contribution to national technological strength.

Table 16.1 Per capita levels of civil R&D for some key economies

Country	GNP per Capita £	Civil R&D per Capita £
USA	14300	379
West Germany	12200	363
Japan	11800	368
UK	10500	170

Table 16.2 is one indicator of R&D output. The US patent applications reflect both scale and strategy in the biggest world market. Among the non-resident applications, Japan dwarfs the rest.

Table 16.2 US patent applications 1992 (Source: Industrial Property Statistics, World Intellectual Property Organization)

United States	94017
Japan	40267
United Kingdom	7040
Germany	14466
France	6012
Total non-resident	93274
Grand total	187291

16.3 EXPECTATIONS FROM GOVERNMENT

It is likely that many expectations in industrial countries are similar and in that respect the United Kingdom furnishes a good example. In the context of industrial renewal and development, the expectations which seemed to have crystallized most strongly from public discussions in the UK during the last two to three years are briefly given below. No ranking is implied in the order and the listing does not assume consensus on every point.

1. A basic philosophy and vision of long-term intent is looked for, within which all aspects of policy can be integrated and updated. The vision must allow adaptive planning for technological opportunities.
2. A stable economic environment is needed which is based on low inflation, low unemployment and low interest rates. This forms the basis for sustainable growth and an economic climate conducive to inward investment by multinational companies.

3. The provision of a more accurate economic forecasting system is very desirable.
4. There is a case for the development of sectorial policy models by the major government departments and their integration within a national perspective.
5. Government departments should further develop R&D partnerships with industry.
6. There should be more emphasis on national economic and technological needs rather than 'preoccupation with the finer points of political philosophy'.
7. Greater effort must be made to maintain and enhance international competitiveness, particularly for the manufacturing industries.
8. The national advanced technology infrastructure must be maintained and further developed. An effective framework of national research services and institutions is deemed essential. These should include the assembly and provision of science and technology information and databases.
9. There has to be continued support of efforts to secure major international contracts. One example of this is the provision of export finance, particularly where it is needed to match overseas project funding.
10. There needs to be greater strategic direction of scientific and technological research based on technological demand articulation, i.e. the systematic search for and selection of technological options. Support is required for R&D in emerging technologies.
11. European and international industrial partnerships should be encouraged, both for R&D and manufacturing.
12. Financial incentives for company R&D should be provided on the basis of defined criteria.
13. A world-class education system is seen as essential with a greater focus on competence and lifetime learning as well as more emphasis on education and training in science and technology.

16.4 NATIONAL OBJECTIVES

It is interesting that many of these expectations are echoed in the recommendations to the UK government by the Advisory Council on Science and Technology (ACOST) (1991).

In order to generate a competitive technology-based economy in the 1990s and the early 21st century, the UK needs:

– a coherent framework of Science and Technology (S&T) goals, strategies and policies, taking into account international develop-

ments, and an adequate and well directed national investment in S&T activities;
– positive public attitudes towards S&T;
– a supportive regulatory regime;
– an effective system of education and training in S&T;
– a vigorous and excellent S&T research base;
– sufficient and effective industrial investment in R&D, in the exploitation of R&D and in innovation more generally;
– an economic climate which encourages innovation, including sufficient economic stability to enable the risks and rewards of innovation to be sensibly evaluated;
– effective mechanisms to identify and exploit emerging and generic technologies;
– effective mechanisms to encourage technology transfer.

16.4.1 Implications

ACOST accepts that it is difficult to separate S&T from other policy areas such as health, the environment, education or defence. S&T issues therefore need to be considered in such other areas of policy formulation.

The direction and investment in R&D can be just as important as the total volume of expenditure. In some countries, such as Japan, a strong sense of direction is provided centrally in areas of potential economic success or scientific excellence. Consider the implications, for instance, of the R&D surveys carried out by the Statistics Bureau of the Japanese Prime Minister's Office where a company's intramural R&D expenditure is disaggregated into different product fields (Kodama, 1991).

There needs to be more discussion about S&T policies aimed at ensuring that industrial, academic and government viewpoints are integrated, that S&T policy studies are disseminated, and common goals agreed. The partners in such discussions should not only be scientists and technologists but should include manufacturing and marketing management.

16.5 THE ROLES OF GOVERNMENT

What can a government do? One first task is to define the mechanisms of cooperation with the parties concerned, such as with industries and universities. How can it meet the stated expectations and achieve its objectives? Its response can be expressed by a number of roles and activities. The following are some of the more important:

1. The supply of stimuli to industry and commerce.
2. The development and implementation of relevant national policies. This is the major initiative role.

3. The provision of support. This can be fiscal, such as tax allowances/ long-term, low interest loans or subsidies to designated industries.
4. To take up the role of facilitator.
5. The use of procurement facilities. Most governments are major buyers of equipment and systems.
6. To provide a focus on markets, such as export opportunities.

16.6 THE SUPPLY OF STIMULI

Porter (1990) succinctly states the role of government as a pusher and challenger; to put pressure on industry to achieve competitive advantage. It can signal opportunities and dangers; identify and highlight priorities and challenges. It can encourage industry by schemes, such as the UK Queens Award for Technology or Exports. The various initiatives of the UK Department of Trade and Industry (DTI), such as the advanced manufacturing technology (AMT) and the manufacturing operations people and systems (MOPS) programmes, express government intent to see industry make use of advanced technology and achieve world class performance. The stimulus can take the form of joint workshops, such as the DTI and Society of Motor Manufacturers and Traders 1993 workshops on 'lean' automotive component supply.

Another good example of signalling opportunities is the publication by the Technology Administration, US Department of Commerce (1990). It identifies 12 emerging technologies and assesses their world market potential in the year 2000. A scenario of competition from the European Union and Japan indicates the challenges in each group of technologies. The purpose of the report is to provide a source of information to be used by US industry, trade unions, government departments and universities, when they develop programmes and policies to exploit these new technologies. This is an important input to a forum for discussion and there is a specific disclaimer that the new technologies have been pre-selected for support. Such a pre-selection strategy may be appropriate for a developing country with limited means, but is regarded as unsuitable for the United States with its large science base and broad spectrum of technological resources. The foreword by the then Secretary of Commerce, Robert A. Mosbacher, illustrates the role of the US Government as a 'pusher' ; the scene is set. It is now for others to take the opportunity.

16.6.1 The focus on the market place

The biggest economic benefit of technological progress often accrues to the businesses which have paid careful attention to market opportunities. The attitude which underlies such a policy is not shared by all companies. Government encouragement and emphasis to bring innovation successfully to the market place is an important facet of a national technology policy.

16.7 THE FRAMEWORK FOR POLICY

16.7.1 The United Kingdom approach

In the UK the nucleus for science and technology policy is the Cabinet Committee on Science and Technology, chaired by the Prime Minister. Industrial and commercial matters, including competition and deregulation are within the purview of the Cabinet Committee on Industrial, Commercial and Consumer Affairs whose chairman is the Chancellor of the Duchy of Lancaster (in other countries the nearest equivalent would be a minister without portfolio). Most of the members are the Secretaries of State of relevant departments.

The Department of Trade and Industry has a primary brief for industry In 1992 the government established the Office for Science and Technology (OST) within the Cabinet Office. In July 1995 the Office of Science and Technology (OST) was transferred to the Department of Trade and Industry and the President of the Board of Trade became the Cabinet Minister for Science and Technology. This move allows the Government's policy on science, engineering and technology to be developed alongside its policies on industry. The functions and remit of OST remain unchanged. The prime role of OST is to develop and oversee the implementation of a strategic vision of the contribution of science and technology to UK life. In 1993 the government published the White Paper 'Realising our potential'. In pursuance of this it formed the Technology Foresight Steering Group charged with the collection of information on scientific opportunities and potential market applications by panels of experts. This includes scanning what happens in all other parts of the world. One objective is to provide early notice of emerging key technologies anticipated within a five to ten year horizon. It is also hoped that this initiative will lead to closer relationships between researchers, industry and government. The group will make annual forecasts. Its main tasks will be to:

- establish the current state of science and technology;
- forecast future scenarios over a five to ten years horizon;
- estimate the likely rate of change;
- transcribe to the national level the technological opportunities perceived by companies.

Its reports will be used for the formulation of national technology policy and government-funded research programmes.

For government-funded research there will be an annual 'forward look' providing details of the portfolio of research projects that would best meet national needs, together with an assessment of how well departmental research programmes match the requirements of this portfolio.

The oversight of this research strategy is the responsibility of the Council for Science and Technology which includes industrialists and customers of publicly funded research. The Council will advise the government on those issues of science and technology judged to be central to the success of the UK. The new Engineering and Physical Sciences Research Council is intended to provide a stronger focus for engineering and technology research.

An important component of the policy framework is the transcription of technology policy into industrial competitiveness. This is a responsibility of the Department of Trade and Industry (DTI). Close relationships with industry are pivotal here. These are achieved by such arrangements as the secondment of industrial managers to help with policy formulation and implementation; links with the Confederation of British Industry (CBI) and various trade associations; and the establishment of business link centres in towns with over 50 000 people.

16.7.2 The Japanese approach

The clarity and long-term nature of Japanese government and business planning was crystallized in 1984 by The Japanese Council for Science and Technology which formulated the following three fundamental objectives for science and technology in the 21st century:

1. to promote the creativeness of science and technology;
2. to seek that science and technology will be much more harmonized with man and society;
3. to strive to promote international activities.

These statements are an expression of the basic view that it is the concern of the state to manage the affairs of the 'nation-family' There is an element of dirigism here, but as Frieman (1987) points out, the strength, of the Japanese system emerges from a unique ability to combine the best features of central planning with the dynamics of market competition in the execution of national R&D programmes. Several different government organizations are charged with promoting national science projects.

The Council for Science and Technology is chaired by the prime minister and is responsible for the establishment of long-range R&D goals.

The Science and Technology Agency (STA), a part of the prime minister's office, frames science policy in terms of national plans, administers very large-scale national science projects and coordinates the activities of other ministries and agencies concerned with science and technology. It controls The Japanese Research and Development Corporation (JRDC).

This body promotes R&D for new technologies by integrating the research efforts of government, academic and the corporate research communities. It also funds certain high risk projects. JRDC operates the exploratory research for advanced technology (ERATO) programme with the objective of stimulating more creative and original research within Japanese corporations. The projects are mission based, have a life of no more than five years and are 'individual centred', with a designated leader having full authority for the project. Typically, a project team contains three to five research groups, each containing about five research staff. These are seconded to the project for a period of two to five years from their employing institutions. A university, company or government laboratory is made available for their use.

The Ministry of International Trade and Industry (MITI). This ministry is directly responsible to the prime minister and in parallel to the activities of the Science and Technology Agency, MITI plays a key role. It has a strong core of top graduates, who are developed as generalists rather than as engineers, and who in their career work closely with different industrial sectors. They act as channels for the promotion of industrial technological policies with the aim of establishing Japan as a 'technology-based' nation. The basic MITI policy is:

1. to provide an environment in which the vitality of the private sector – the foundation of national technological development – can be maximized;
2. to promote development programmes in organic linkage between industries, universities and government in those areas where development costs, the terms of research and inherent risks, make it difficult for private enterprise to carry out these development programmes without assistance.

Through the provision of 'seed money' MITI funds basic research projects and contract research to solve particular problems. It operates mainly through its Agency of Industrial Science and Technology (AIST) and supports both joint laboratories and corporate laboratories.

AIST supports important technological developments in the private sector through tax incentives for technological development. Research projects are selected, following an opinion gathering exercise to obtain a consensus in which MITI plays a coordinating role. Very close contact is maintained during the research project with the companies who receive MITI support. This provides MITI with a detailed awareness of the state-of-the-art and technical bottlenecks which call for specific funding.

The Key Technology Centre, set up jointly by MITI and the Japanese Ministry of Posts and Communications, provides loans for company R&D and capital for special R&D companies. According to Engel (1987), it arranges surveys of key technologies, facilitates cooperation between industry and government laboratories and enables foreign researchers to work in Japanese companies. Significantly, it allows a R&D company established under its aegis to have foreign equity participation.

The Institute of Physical and Chemical Research (RIKEN) operates a 'frontier research programme' where long-term research projects are undertaken with an horizon of up to 15 years. According to Engel (1987), the research will be carried out with a substantial proportion of foreign staff and team leaders.

Taking an overall view, the Japanese framework for the formulation and implementation of science and technology policy is indeed impressive. Considerable attention is also given to the utilization of its outputs. An example of this is the following remit of the Technology Promotion Division of the Research and Development Utilization Office.

1. to assist private sectors financially in providing them with conditional loans for their R&D activities;
2. to foster industrialization ... by recommending qualifying research items to the Japanese Development Bank for low-interest plant and equipment loans;
3. to assist manufacturing research organizations ... by preferential tax treatment.

16.8 THE FORMULATION OF NATIONAL TECHNOLOGY POLICY

There is a distinction between a centrally defined national strategy and the encouragement of strategies at sector or industrial levels. Some economies, such as the German and Japanese, are thought to have gained from the participation by government in the development of industrial strategies. This has been achieved through the promotion of strategic awareness and tactical thinking by companies in a specific sector.

Much derives from underlying national philosophies. There is a significant difference between those countries which have a clear, long-term strategy towards the development and exploitation of specific advanced technology product groups and new technologies, and those which do not.

For instance, the French approach to advanced technology is expressed within the three to five year national economic plans for the French economy. The government controls over 50% of national industrial R&D. Through its 'mobilization programmes' it supports five major industrial areas: steel, chemicals, electronics, health and materials. Each industry is viewed as a 'filière' which expresses a chain of production from raw material to the final product.

16.8.1 The objectives of a national technology policy

One major purpose of a national technology policy is the harnessing of technology to meet economic and social goals. It needs to encourage the deployment of national resources with a high and rising level of productivity, leading to the upgrading of the economy and industrial capability.

In essence, technology policy, as a constituent of overall economic policy, is ultimately concerned with the creation of wealth.

Rothwell and Zegveld (1985) argue that a prime aim of technology policy must be to help overcome the technological, institutional and cultural constraints that impede technological progress. Government support is needed for the development of new generic technologies which create new industries. To this end it has to influence particular industrial sectors and activities.

Revelle (1975), writing in the context of US science and technology policy, considers that such a policy needs to have three major goals:

1. to maintain the health and effectiveness of scientific research and technological development;
2. to assure the maximum usefulness of scientific and technological advances in serving the people's interests;
3. to provide for the evaluation and assessment of unforeseen or undesirable effects of new technology.

16.8.2 Sectorial strategies

Within the framework of an overall strategy there is the argument for particular attention to certain industrial sectors because of their role in the national economy. Two main reasons are generally advanced:

1. they provide key generic technologies, such as information technology;
2. they are basic to the industrial infrastructure.

One expression of a sectorial strategy was the Application of Computers to Manufacturing Engineering (ACME) Directorate of the former Science and Engineering Research Council (SERC). On the basis that manufacturing industry is essential to the prosperity of the UK and that higher education institutes (HEI) have an active part to play in supporting manufacturing companies, it formulated the following mission statement in 1991 (ACME, 1991):

> to promote, at higher education institutes, highly innovative research and training in advanced manufacturing and its management, closely integrated with the needs of industry.

(a) Problems with sectorial policies

Some industrial sectors are too heterogeneous to be meaningful for a uniform approach to policy development. They often reflect a history of industrial organization, set up originally for other purposes, and do not match the context for technology policy. Technological opportunities are unlikely to match sectorial boundaries.

Policy makers need to consider the technological links between different branches of industry. In some cases, therefore, sector-specific policies may be inappropriate and a balanced, coordinated development programme may be more suitable for technologically-linked sectors. Policy options need to be based on techno-economic constraint and opportunity analysis.

16.8.3 Technological innovation policy

As part of a longer-term regeneration of national competitiveness, innovation is an important part of technology policy. Its realization will require the fusion of certain economic, industrial, science and technology policies. The following will be involved:

- Fiscal policy – taxation and allowances, such as extra tax credits for R&D expenditure.
- Financial policy – grants, such as support for pre-competitive research programmes or development support for small high technology firms.
- Educational policy – providing the technological capability.
- The redevelopment of the supporting infrastructure – the framework of encouragement and direction.
- Technology transfer initiatives.

Tassey (1986) indicates that one of the challenges in developing an innovation policy is that industrial technologies are not homogeneous. Their constituent elements differ in many ways which have economic implications. Some are highly proprietary, and if they yield a high rate of private return, they are well developed. Others pose complex research problems, call for scarce specialist staff and require heavy investments. The returns become unattractive and this leads to industrial under-investment. Despite this, some of those non-proprietary elements may be critical for the evolution and subsequent applications of certain generic technologies. This makes the case for government support of those elements of a national technological base which leverage the proprietary elements of the industrial technologies.

Another aspect, pointed out by Abernathy (1978), is that the nature of technological innovation takes on different characteristics as a productive unit progresses from an initial, fluid stage to a well-established, specific form. There is, therefore, no one best way for government policy to stimulate technological innovation. The appropriate types of stimuli, the coupling of scientific advance with innovation and the barriers to innovation all vary with the different stages of technology development. The concept of innovation as a linear process can therefore mislead the selection of policies that would be appropriate to encourage innovation at different stages.

While the context of innovation policy is primarily national, the regional aspects are becoming increasingly important. A good example here is the European Union. For instance, Bangemann (1992), in his role as European Commissioner in charge of the internal market, emphasizes the importance of a European industrial policy with the main aim of making the European union internationally competitive. He is pragmatic how this should be done. To achieve this, in his view, industrial policy cannot be completely consistent, sometimes you cannot leave it to market forces and state support may be justified. He comes to the interesting conclusion that there is a case for making decisions behind closed doors and relatively quickly. Too much scrutiny and public debate amounts to 'public gauntlet running'. An homogeneous European home market needs to be established so that European companies can grow and achieve the critical mass necessary for effective global competition. What is critical mass varies with time and industry.

(a) The development of infratechnologies

The availability and efficiency of an infrastructure such as roads, railways, telecommunications, water and energy supplies is essential to national competitiveness. ACOST (1991) notes the opportunities for the utilization of advanced technology in the further development and operation of such systems, such as computerized control of urban traffic, the remote control and monitoring of gas and water supplies. Apart from improving the effectiveness of the infrastructure there are opportunities for the development of products and systems which can also be used for the development of infrastructures in other countries.

Quite apart from a sound technological infrastructure to provide for the current needs of industry, technological innovation, in turn, calls for the further development of infratechnologies. For instance, the provision of advanced measurement technologies, such as nanotechnology, testing procedures, integrated within standards or related databanks, provide support for the industrial innovation process. The pool of such methods and data generally do not provide individual companies with proprietary advantage, but are nevertheless used by companies in their operations. They have a risk reduction character which lowers transaction costs. Often, they are competitively independent and can be developed on a collaborative basis with government laboratories. Such technologies are essentially a public good in the same way as a public transport system. It is part of a coherent national technology policy that these are fostered. Furthermore, if the developed technology becomes incorporated in an international standard it gives a competitive advantage to the originating country.

(b) The encouragement of technology transfer

One factor underlying a high rate of productivity growth is the rapid diffusion of innovation, particularly through investment in advanced manufacturing technologies. There are various reasons why the adoption of advanced technologies has been slower in Western countries than was first expected. The encouragement of the required technology transfer, especially to small/medium enterprises is an important ingredient of technological innovation policy.

A report in 1991 by the Organization for Economic Co-operation and Development (OECD) on technology in a changing world considers the capability of an organization to utilize external technological knowledge. The assimilation of such knowledge requires prior knowledge, i.e. technological advance is a cumulative process. Learning by doing, using, selling is an interactive process. OECD stresses the need to give attention to the actual diffusion process in the design of diffusion policies.

(c) Continued participation in fundamental research

Because of the high risks and costs involved in the strategic development of some new technologies, which individual companies are unable to finance, it is argued that government should continue to contribute to selected projects. The most useful areas would be the 'technology drivers' i.e. key technologies which will become multipliers of other technological developments. Even if firms could fully support such expenditures there is the problem in commercial terms of low appropriability of the benefits of such work. To a government this adds to the difficulties of choice, because of other research fields where it might have to provide most of the support, such as in the areas of environment, education or health care.

(d) The development of research and training institutes

Institutes, such as the Fraunhofer organization in Germany, which combine contract research, the postgraduate training of engineers and applied scientists, with strategic research in their own industry or technologies, have made major contributions to the technological strength of their country. The emphasis is on emerging technologies and applied rather than basic research. The diffusion of generic technologies is fostered by close links with prominent university departments and industrial companies. Often the director of a local research unit is a full professor at a nearby university. There is much in favour of such institutes in the context of balanced industrial science and technology strategies.

(e) Support for high technology industry

A good illustration of this aspect of innovation policy is provided by Porter (1990). He considers that the most important role of Japanese cooperative research is to signal the importance of emerging technological areas and to stimulate proprietary company research. Much of this is undertaken by 'high technology industry'. This is interpreted by the Bureau of Economic Analysis, US Department of Commerce as an industrial sector which satisfies one of two conditions:

1. the R&D expenditure within the sector is at least 10% of its added value;
2. scientists and engineers exceed 10% of the employees in the sector.

Dosi (1984) considers that the basic Japanese policy here is to remove structural constraints which impede commercial exploitation by:

1. initial formal and informal discouragement of direct foreign investment;
2. public definition and monitoring of the terms of licensing agreements to benefit an industry not just a company. This allows automatic diffusion to all Japanese companies.
3. initial import controls;
4. setting of technological targets.

Japanese industrial and trade policies allow a definition of company strategy consistent with national objectives, making it rewarding for a company what is considered necessary for the national economy. Consistent and comprehensive national research and development policies are an important factor in Japan's level of national performance.

(f) The fusion of technologies: a case illustration

An example of the relevant Japanese industrial policy is the Law for the Development of Specified Machinery and Electronic Industries (1971). The law encourages reciprocal research investments between industries with special low-interest loans. For instance, the ceramics industry concentrates on applications for electrical equipment while the electrical industry undertakes complementary research in ceramics. As indicated by Kodama (1991), the partner industries integrate their research/technology strategies. Similarly, the subject of mechatronics emerged through reciprocal research investments in the fields of machinery, precision instruments, electrical equipment and communications/electronics.

The consequences of this emphasis on high technology are indicated in the 1991 OECD Report *Technology in a Changing World*. During the period 1970–85 the growth rate for low technology manufacture in Japan was about average for OECD countries. For high technology, the corresponding growth rate was about 14% above this international average. The compound effect over 15 years has to be appreciated.

(g) Emphasis on research intelligence

Another key component of technological innovation policy is the emphasis on research intelligence. Again, Japan provides a telling example. Frieman (1987) points to the careful and systematic scrutiny of foreign research work and results. This includes the full text translations of foreign technical journals and reporting on international scientific conferences, the maintenance of comprehensive scientific databases, the establishment of joint ventures with foreign corporations, obtaining technology licences and assigning research staff to carry out concentrated studies of foreign technology. Significantly, the results of foreign R&D, where appropriate, are incorporated in long-term national and corporate business development plans.

McMillan (1985) points to the role of government agencies such as the surveys conducted by the Japanese External Trade Organization (JETRO), an affiliate of MITI, which draws economic and technological intelligence from over 100 offices in major trade centres.

16.8.4 The retention of a national technological capability

An important policy objective, particularly in a period of recession, is the maintenance of technological capability. Industrial management is under pressure to reduce both direct costs and overheads, to slim down or even break up research and design teams. But once disbanded, these take time and costs to reconstitute and the chance to catch up in advanced technology is small. There is a further and deeper aspect emphasized by Tassety (1992). Long-term economic growth is a function of the total added value by an economy. In turn, this is affected by knowledge production; i.e. the location and scale of R&D within that economy. One way to impair national competitiveness is to cease research and development work; to become hostage to the technology strategy of others.

Government support for research and development of new technologies, technology demonstrators and new equipment programmes is an important contribution to the retention of technological capability. It can be effected through government-linked research agencies, research associations and universities. The UK link and teaching company schemes which, in essence, are consortia of government, universities and industry also have this basic objective. For example, one of the first initiatives by the link scheme in 1988 was the nanotechnology programme, concerned with ultra-precision measurement and manufacturing processes. So far 20 projects have been approved to the total value of £23.5m. The 50% funding is for collaborative research programmes involving industry and universities.

Where the product life cycle is relatively short, because of the rapid substitution by new products, the rate of change both in product and process technology is often high. The cumulative nature of innovation provides momentum to a country and production is less likely to flow to imitating countries.

16.9 THE CHALLENGES TO POLICY

Omniscience is not a characteristic of government. The limitations of central governments have to be appreciated. Where governments are 'democratic' they are influenced by political considerations. For technology policy development, key staff with practical industrial and commercial experience are not always available and a credibility gap can easily develop between central government departments and their corresponding industrial sectors. The first challenge is the very structure of policy making.

16.9.1 Features of a coherent policy

The formulation of technology policy requires, in the first place, a sound basis and professional standards for policy analysis. This includes preparatory investigations, assessments, the formulation, integration and reconciliation of potential policies and the systematic evaluation of existing policies. In turn, effective policy implementation calls for organization, training and the development of infrastructures. Such activities require planning, funding and quality control.

Rothwell and Zegveld (1985) conclude that, whatever the substance of a technology policy, it needs to be coherent and to be focused. This means that the various institutions involved in the formulation and implementation of policy integrate their activities so that they do not contradict each other, nor are misaligned with other national policies. There has to be consistency, which is above the cut and thrust of parliamentary politics. The time-scale for structural development requires consensus on a longer term basis.

Yet despite this, flexibility is important so that policies can respond to changing industrial needs, threats and opportunities. National policies should be in harmony with the strategic interests of individual companies. Last, but not least, is the requirement of realism; the awareness of what public policy can and cannot do.

16.9.2 Conflicting goals

Many of the values and goals of a pluralistic society are inconsistent and conflicting. This multiplies the challenges to national policy making. There is also the dilemma between the provision and the distribution of private and public goods. While there may be agreement in principle about the overall quality of life, there is no agreed means of measuring this. The Gross National Product (GNP) *per capita* is an imperfect guide.

16.9.3 Limited systematic data

It is important that systematic statistical data are available for policy makers, as otherwise policy could be reduced to responses to pressure groups and be more short-term than it should be. Strategy needs a basis. For instance, more effective indicators are needed for the assessment of the benefits of industrial investment in R&D. A distinction can be made between research expenditures and development expenditures. At the level of the firm, these expenditures can be subject to different accounting principles. As we have already seen in Chapter 10, they often reflect different activities. To aggregate these into one macro-indicator, as at present, does not help the judgement of policy makers. It is interesting here that OECD (1991) also recommends key indicators based on broader concepts of innovation and diffusion.

There is also a case for additional social indicators which reflect the quality of life. For instance, product innovation with certain consumer goods does not show up well in GNP measures; e.g. a power tool for DIY (do-it-yourself) can substantially increase productivity in the home, but this is not measured. If there is a resultant drop in the income of a trades-man an innovation could even lead to the reduction of the GNP!

16.9.4 Problems of technology assessment

It is important that policy advisors and key staff in government depart-ments have the capability to assess the quality and productivity of science and technology programmes, particularly if science and technology pro-jects are approached in investment terms rather than as annual work programmes. Furthermore, policy makers are not always familiar with the industrial innovation process and may be tempted to think more of R&D aspects.

Also, the limited market expertise of policy makers could result in the funding of technologically sophisticated projects with limited market scope and profitability. The selection of advanced technology projects with high market potential is a particular challenge for government policy makers. The effective sifting and audit of advice is crucial.

As assessment is normally a precursor of choice, it is also important not to overlook that:

• The choice of specific options assumes the existence of a manageable context and that outcomes are predictable.
• Choice presumes a basis of objective information.

16.9.5 The need for standard administrative practices

The final report of the *Alvey Programme* (1988) indicates the difficulties with some government-supported initiatives where considerable time is lost in developing an administrative framework, resulting in a duplication of effort and not enough speed of response. The lack of administrative procedures during the early stages could have been alleviated by a preparatory working group between the announcement and start of a programme. A standard administrative practice for all initiatives, irrespective of government department, would be useful. The analogy with industrial R&D, with lead time cuts and concurrent engineering is relevant here.

16.9.6 The speed of response

While technology can change rapidly, political and social institutions, certainly in the West, change relatively slowly. Furthermore, they seldom shift in anticipation of the social needs arising from technological change but only in response to the resultant pressures. The problems have to be manifest and in a democratic society the debate can be prolonged.

16.9.7 Limits to policy

Because of the nature of technological innovation, with its variations between industries, firms, managements and the very technologies themselves, there are inherent limitations to what an innovation policy can achieve. Governments can furnish the enabling conditions to encourage progressive management. It can do little where managers are indifferent or incompetent.

16.10 THE PROVISION OF FISCAL SUPPORT

In some countries the implementation of national technology policies is fostered by fiscal measures. In essence, these are either tax concessions or grants. The tax concessions come mainly in the form of extra allowances for approved expenditures. These can be on capital account for the purchase of equipment or on revenue account such as for R&D expenditures. Grants may be linked to certain schemes, as with factory developments in UK development areas or where the government becomes a research partner in a particular project on a predetermined funding basis. Generally, grants are related to specific proposals whereas tax concessions usually apply to categories of approved expenditures.

16.10.1 Tax concessions and technological objectives

The use of fiscal policy to support specific innovations is best illustrated by a Japanese example. As an instrument of Japanese economic and technology policy, taxation and depreciation incentives are provided, for example, if certain types of equipment are bought. During the period 1977–82, purchasers of CNC machine tools with less than 60 Kbits of memory and accuracies up to 0.005 mm were accorded a special depreciation incentive of 13% of purchase price in the first year in addition to ordinary depreciation. The purpose of this incentive was to help the key machine tool industry and the way it was implemented was to help its customers. This support took a new form in 1985–89 when, within the theme of investment in new technologies, companies, with under 1000 employees, qualified for special incentives when buying CNC machine tools with accuracies up to 0.001 mm and simultaneous numerical control of at least two axes. They obtained a tax deduction of 7% of the purchase price in addition to a special depreciation allowance of 30% in the first fiscal year in addition to ordinary depreciation.

The scale and directness of aid, which focused on medium-sized companies, and the technical requirements of support are telling. During that period the annual output of CNC lathes produced by Japanese companies increased from about 1300 to 12000. Japan now leads the world in CNC machine tool production. It is estimated that in 1990 Fanuc, the leading Japanese CNC producer held about three-quarters of the domestic market and about half of the world market.

16.10.2 Company taxation

A common form of fiscal support is to allow a company 'free' equipment depreciation; i.e. the company can deduct depreciation against taxable profits to any time pattern it wishes. An extended form of support is to provide tax allowances in excess of equipment purchase costs. Allowances can differ with the type of equipment purchased; the company car need not be in the same category as a machine tool. Also depreciation policies can vary geographically such as with particular 'development areas'. One of the problems of such support can be its lack of focus, although the previous example showed how this was overcome in Japan.

We have already noted in Chapter 10 that a company normally debits research expenditures against its profit and loss account. In that respect the expenditure already qualifies for tax relief. It is a matter of public debate whether the tax relief should be higher than the incurred expenditure. For instance, the report *Innovation in Manufacturing Industry* by the UK House of Lords Select Committee on Science and Technology (Session 1990/1) recommends that 150% of corporate R&D

expenditure should be exempt from corporation tax. This would be based on the amount of real additional expenditure a company makes over its previous year total.

Unless targeted with care, tax incentives can be a blunt tool. For instance, there can be problems with the measurement and apportionment of R&D expenditures. A tax incentive is a help only against tax liability. If a firm makes losses, and this is not unusual with a new, high technology firm trying to get a foothold in a market, the incentive is all but lost.

16.10.3 Risk sharing with companies

Another form of support is that adopted by the state-owned Bank of Communication of Taiwan. The bank provides risk capital to companies which plan to produce a product classified as strategic and which can satisfy a rigorous screening process. To be classified as strategic the product will have to have high added value; high linkage effects; a high skill content and a low energy intensity. The evaluation comprises both financial and technological assessments. It is argued that a small group of carefully chosen officials have better information than many a small or medium company. This enables them to pick winners better and to allocate scarce capital. Capital is provided at 2% below the prevailing market rates. In essence, the bank provides a subsidy by supplying risk capital without the usual risk premium.

Where a company obtains total financial project support the risks are shouldered by the government and in some instances marginal projects could be pursued because of this financial cover. The belief in the success of a new technology venture is best demonstrated by some company money invested in it. An alternative approach is based on gradually increasing company contributions as a project progresses and the technological risks diminish.

16.10.4 Tariffs

These are general instruments of policy and their overall effect again is somewhat blunt. For developing countries they are sometimes attractive because they provide shelter for infant industries. In the long run they lead to distortion and should the sheltered industries be exposed at any time to world competition then they quickly become vulnerable. For developed nations, particularly with heterogeneous industries, they are relatively inefficient. An infant industry can best justify protection on the basis of proposed strategic changes. In that case the nature and duration of the tariff should allow for the technology gap and the required development time of the industry.

16.11 THE GOVERNMENT AS A FACILITATOR

A facilitator is someone who helps the affairs of others. A government can do this in a number of ways with or without financial support. For instance, the 'Enterprise Initiative' of the UK Department of Trade and Industry provides a package of advice, guidance and practical support for British industry. This includes, for example, a practical guide for technology-based businesses on getting the right financial support. There is advice on collaborative research and help with finding project partners. More effective management of computer-integrated manufacture is fostered by publications, seminars, case studies, workshops, videos, mobile demonstration units and visits programmes.

The Technology Administration, US Department of Commerce in its survey of the technical and economic opportunities provided by the emerging technologies (1990), emphasizes that the general efficiency in the use of technology depends on the availability of generic know-how, information and facilities. It recommends government participation in industrial consortia, the ready availability of methods, references and data. It was also important to give industry access to government facilities, and to participate in international standardization activities to ensure that the interests of domestic industry were safeguarded.

16.11.1 Reinforcement of technological resources

One very important form of assistance, particularly for small- or medium-sized companies, is the provision of consultancy support for the evaluation of technological projects, such as automation schemes. The lack of appropriate previous experience and a simplistic approach can already at the very early stages of a project be the source of excess costs, delay and risk of failure. A rigorous feasibility study carried out with an established consultant is an important first step for many companies with limited resources. It must also be remembered that a feasibility study is not an automatic endorsement to proceed and that, with a given business context, automation is not always the way forward.

16.11.2 Pre-competitive collaborative research programmes

The complexities and heavy research costs of some emerging technologies have brought about increasing government involvement as initiators and facilitators of industrial research consortia. These consortia include the major firms and competitors in a given industrial sector who are persuaded by risks, costs and encouragement to collaborate in the early stages of technology development. The case for such consortia has been strongly argued by Tassey (1992) who points to:

- the economies of scope which exist and which, for a variety of reasons, may not be taken up by individual firms;
- the very complexity of the research process, which calls for a wide range of skills and facilities;
- the economy of scale of research processes;
- the stimuli and the learning opportunities provided by the mingling of different company staff.

The formative influence on commercial companies, working with universities and rival firms on collaborative pre-competitive research, provides a foretaste of working together which may be extended to other activities. Participation in the formulation and implementation of such programmes brings companies closer to the policy-making core and increases the chances of influencing future government strategy.

The technology transfer of research accomplishments for the application developments by individual firms is an important function of such research consortia.

(a) An example of a pre-competitive research programme

As part of the Advanced Technology Programme, supported by the UK Department of Trade and Industry (DTI) and the Engineering and Physical Sciences Research Council (EPSRC), there is the research workplan for safety-critical systems, within the Joint Framework for Information Technology (JFIT).

JFIT was established in 1988 to extend coordination between DTI and EPSERC, each of which support information technology (IT) but are accountable to different Departments of State. It aims to create coherent policies, objectives and priorities for IT research activities in universities, companies and national research institutions. It provides a framework, furnishes guidelines and integrates the activities of independent organizations.

The coordination of policies is carried out by an Information Technology Advisory Board (ITAB) which provides advice on strategic policy issues, research priorities and resource allocation. The responsibility for specific policies and programmes is delegated to specialist advisory boards, such as for devices, systems architecture etc. The members of these boards are drawn from industry and the academic sector. They include both users and suppliers of IT.

The Advanced Technology Programme is designed to encourage pre-competitive research in new technologies and to accelerate their application. Furthermore, it aims to encourage smaller firms to participate in IT research, strengthen the links between industry and the universities and involve the users of IT. All this is with international competitive advantage in mind. As Tassey so aptly puts it: it is better for an industry to be competitive than to have no players at all.

The Safety-Critical Systems Programme reflects the impact and implications of technological change. There is the widespread introduction of computer-based control systems to safety critical applications, e.g. aircraft or nuclear reactors. Computers coupled with advanced actuators can now provide cheap, compact, but sophisticated systems. The programme will be judged mainly by the adoption of new techniques. To this end the applications developed within its remit are expected to be realistic demonstrators, addressing downstream as well as horizontal technology transfer. The programme requires that the results of such demonstrators can be quantified in commercial terms and that they have been obtained in a shop-floor environment.

An interesting feature of the programme is the establishment of a safety-critical software club to ensure that each project takes into account parallel work and contributes to the overall programme aims.

The three main research themes of the programme are 'technologies', 'human factors' and 'unification'. The last theme is concerned with the development of a common basis for the comparison of alternative sectoral and technological approaches and particularly seeks coherent practice within sectors, with the integration of a range of procedures and techniques applied to a range of technologies.

On the basis of public funds for up to 50% of the total eligible cost, there is a substantial schedule of project requirements.

16.11.3 Support for small- and medium-sized enterprises (SMEs)

Broadly speaking, SMEs are firms up to 500 employees. They can be divided into those which are in traditional industries, those which are modern and with a niche market and new technology companies. The latter group, with its promise, warrants particular attention.

The hallmarks of such firms are limited financial and human resources and little economy of scale. The SMEs in traditional industries are often unable to identify and use external expertise and advice. They are not familiar with the 'corridors of power' and less skilled in persuading government to provide financial support and research grants. There is also the reluctance to take key staff out of day-to-day operations and to harness the opportunities provided by technological progress. They have limited risk-taking capability.

One important form of support for such firms is the establishment of a network of regional groups to bring together manufacturing companies and university departments to disseminate best practice in such fields as advanced manufacturing technology or concurrent engineering. This has been further developed by the UK Department of Trade and Industry (DTI) with the provision of innovation and technology counsellors through its business links scheme. These counsellors will provide technological information and guidance with the help of national data networks.

The needs and scope for support of new technology firms are somewhat different. Some new and rapidly developing knowledge based technologies, such as information and biotechnology, have provided opportunities for entrepreneurial specialists to set up businesses. Knowhow is crucial and entry costs are modest. These firms are the seedbed for future industries and therefore warrant government attention. Rothwell and Zegveld (1985) enumerate the scope for government to encourage the start up and development of such enterprises. The main features of such support could typically be:

1. The provision of venture capital or loan guarantees for such funding. This is critical where orders are dependent on further product development;
2. The adoption of risk-accepting public procurement practices;
3. The encouragement of university innovation centres which provide a focus for local entrepreneurs;
4. The improvement of personal mobility by the modification of pension schemes.

One of the great risks of an economic recession can be the demise of young new technology companies which do not have the accumulated levels of reserves, experience and contacts. While the long-term prospect may look good, lack of orders and cash flow problems pose an immediate threat.

16.12 THE INSTRUMENT OF GOVERNMENT PROCUREMENT

Public procurement by government departments, such as the UK Ministry of Defence, has reduced the risks of market entry for small new technology firms. Their products are often components or subsystems of larger installations and have other, commercial applications. An airborne military camera is such a case. Much depends, of course, on a department's strength as a buyer. Where it is a substantial or the whole purchaser its influence on the technological development of the company's products is correspondingly enhanced. Similarly, where the product design is at an early stage of development, government procurement can more effectively encourage technological advance.

The manner of procurement can encourage innovation in other ways. For instance, a government department can purchase a limited number of prototype units on a competitive basis. It could also place a development contract to establish the feasibility of an innovative product, to be followed by manufacturing orders. Various incentive or cost-sharing devices can be applied. Life cycle costing is also used where an innovation indicates lower operating costs which more than compensate for an initial higher bid.

One aspect, particularly with defence purchases, has been the development of procurement by 'cardinal points specification'. Cardinal points are essentially concerned with the key performance requirements of a product. How a supplier achieves these is a matter for his creativity in design and his ability to use the latest technologies. While this approach has a number of commercial considerations it also gives more scope for technological development. In the past, fixed technical specifications discouraged innovations by suppliers. An important additional feature is the government ability to provide proving grounds and test facilities for prototypes.

Porter (1990) observes that governments can upgrade national competitive advantage by the early procurement for advanced products; being demanding and sophisticated buyers with strict product specifications and not just accepting what suppliers offer. This is important for product performance, safe operation and environmental requirements.

16.13 THE REGULATORY ENVIRONMENT

In a responsible modern society, regulations to secure environmental, health or safety improvements are important. However, over regulation can stifle innovation and an appropriate balance needs to be struck. Schedules of compliance with regulations need to be established sufficiently early so that adequate warning is given for their effective incorporation in product design. Uncertainty about regulations and their implementation can discourage firms to invest in high-risk innovation.

Also, regulations are more meaningful and acceptable where industrialists and other experts participate in their formulation. Where industrial involvement is initiated at an early stage there is a fuller understanding of the problems of the other parties concerned.

A distinction can be drawn between the content and the administration of regulatory standards. There is a collective gain where standards are established promptly, applied consistently and managed efficiently. International equity in enforcement practice is important to avoid being disadvantaged because of less rigorous enforcement by competitor countries.

16.14 THE INTERNATIONAL CONTEXT FOR TECHNOLOGY POLICY

16.14.1 The concept of a niche economy

It is difficult even for the largest and richest economies to supply all their industrial needs and to be world leaders in every branch of technology.

The inherent limitations of economic and human resources force a country to choose its focus and concentrate on a limited number of key industries and technologies which can achieve and maintain world class capability and status. The aim is to establish strategic 'clusters of excellence'. The challenge is which industries so to designate and develop.

16.14.2 Added value in international trade

One indicator of a country's technological state of health is the added value of its volume of export compared to that of its imports. This can be achieved in two basic ways. The first is the enhancement of established products where premium quality and enhanced product attributes generate greater value and therefore command higher prices. The other way is the development of innovative products which open new markets.

Where the international transfer of technology is 'naked', such as with the licence of a patent, rather than embodied in equipment or consumer goods, then at the macro-level the value of 'displaced' exports and the resulting imports can be much greater than the value of the fees received and the exports of the product from the country of origin. This is particularly so where the arrangements make the licensed technology available to all appropriate firms in the country – as has happened in Japan – and this is coupled with rapid further technology development.

16.14.3 The integration of UK and European Union policies

With the advent of the single European Union market, a discussion of national technology policy in Europe would be incomplete without reference to this major development. The growing interlocking relationships between the European Union (formerly the European Community) and its constituent countries have now become a significant factor in national policy. The single market has changed the economic and industrial context of Western Europe. It brings opportunities and uncertainties.

The first major opportunity is the very size of the new market with practically unhindered access to a population of about 370 million. Economies of scale will reduce manufacturing costs and allow European firms to compete more effectively with imported mass produced consumer goods. It will also give greater scope to smaller, specialist manufacturers.

The second important opportunity is the integration of key research and development resources so as to make better use of high calibre staff. To quote the Commission of the European Communities (1991):

> Europe does not lack potential, it has many research scientists, mostly of extremely high calibre. Its research spending, while lower

than that of its competitors, is still quite respectable. Nevertheless, divided as it is into different countries, Europe does not make the most of its substantial intellectual and research potential. Dispersed funding, isolated research, lack of mobility among research scientists, duplication of effort under national programmes: these are all factors which tend to reduce the impact of European research work.

It is not the policy of the Union to transfer as many research activities as possible to the European level. The basic aim is to promote projects only at the European level if, because of their size, cost and complexity, they would be less effective to run or not be affordable at national level. This applies especially where large-scale research and technical development (RTD) activities, such as controlled thermonuclear fusion, require a critical mass of resources. As specified in the Community Third Framework Programme 1990–94, it concentrates on very major initiatives, such as information and communication technologies, advanced materials technology, biotechnology and global warming. The Fourth Framework Programme for the period 1994–98 will continue with much of this work, particularly with research in information technology, telecommunications and industrial technologies, such as design, engineering, production systems and human-centred management. An interesting development is the evaluation of science and technology policy options and technology assessment.

The implementation of the main Framework Programmes is through initiatives, such as EUREKA, SPRINT and BRITE. The EUREKA initiative (administered by the European Research and Coordination Agency) consists of the European Union states as well as the EFTA countries and Turkey. It supports essentially application projects closer to the market place. SPRINT, the strategic programme for innovation and technology transfer is particularly concerned with the dissemination of new technologies and innovations by integrating national innovation structures into a European network. The insistence on international partnerships is a fundamental feature of all European Union projects. BRITE (basic research in industrial technology in Europe) concentrates on collaborative industrial research projects which must involve at least two independent industrial companies from different member states. Each pre-competitive research project is expected to make a significant industrial and economic impact within two to three years after completion.

Although the strength of political association is a matter of debate, there is a basic government commitment to the European Union. It is now a matter of participation in the economic progress of and the technology growth within the Union. This includes participation in the development of 'trans-European networks' such as the telecommunications and transport infrastructures.

16.14.4 The international harmonization of standards

An important point made by the Advisory Council on Science and Technology (ACOST) (1991) is that the international harmonization of standards, testing and certification procedures is central to removing technical barriers to trade. The achievement of the free movement of goods is also basic to the single European market. To avoid the diversity of product safety laws within the European Union becoming a new barrier to trade, Community Directive 83/189/EEC has established a procedure for the provision of information in the field of technical standards and regulations. The involvement by UK companies and trade associations in regional (EC) and international standards making can lead to a stronger competitive position.

16.15 EDUCATION AND TRAINING

The overall field of education and training is one of the major policy challenges for a developed country. It is too big to be developed here, but not to mention aspects relating to technological competitiveness would leave this chapter incomplete.

For a competitive industrial nation sound general education for all citizens is a first requirement. Science and technology in general education is envisaged to provide part of the intellectual training for informed citizens and prospective leaders. An informed, positive outlook is required towards technological achievement, irrespective of where in the world it was accomplished. Following this, the enhancement of the relevant competence of industrial personnel in key sectors, at all levels, is an important requisite for the required human resource level to sustain growth and international competitiveness. To this end Porter (1990) stresses the need for specialized skills in specific fields. It is particularly important to have well trained engineers and research workers.

Without an appropriate education and training policy there will be insufficient skill and expertise to create, design and manufacture world class products. The capacity to turn complex ideas into high-value products will be impaired and manufacture will be increasingly restricted to assembly operations, using at best semi-skilled labour.

To support and develop the competitive position of particular industries government support also could be 'function-specific'. For instance, it could subsidize the specific training, say, of design engineers in a particular field where there is a crucial shortage. Support can also be 'task-specific', such as the improvement of decision skills in the field of technology management.

16.16 THE EVALUATION OF TECHNOLOGY POLICY

This chapter has been concerned with the need for, the nature of, and the difficulties confronting national technology policies. There remains the judgement of their worth.

Rothwell and Zegveld (1985) have given a comprehensive account of the nature and the problems of policy evaluation. The very first challenge is the multi-dimensional nature of national policy. What are the goals? What are the priorities? If you pursue this policy, what are the effects on other policies? How is government and national income to be distributed? The decisions have a political dimension and this involves value judgements. How do you judge whether the money is well ispent? and whose perception is to be taken?

To provide any form of answer at all, some framework for analysis is needed, such as the type put forward by Gibbons (1982). In his classification three strands emerge:

1. The allocation of resources to science and technology.
2. The organization and control of economic activity.
3. Public policy analysis.

The third strand, according to Nelson (1974) is concerned with the laying out of alternative courses of action, of tracing their consequences, in terms of costs and benefit, and identifying at best or at least a 'good' policy. The utility of such an analysis is seen in providing guidance for the policy maker.

Some of the major hazards of evaluating technology policy have been indicated by De Leon (1976) in his assessment of nuclear development strategies. He concludes that major technology development programmes can best be viewed as having multiple actors and objectives. This means that a single evaluative standard for a technology development programme is inadequate and potentially misleading. There is a need, therefore, for multiple success indicators, for comparative studies and for evaluation criteria suited to the different outcomes expected from the various stages of the development process.

16.16.1 Technology assessment in the public domain

As pointed out by Berg (1981), technology assessment is a class of policy studies which examines the effects on society when a technology is introduced, extended or modified with special emphasis on unintended and indirect consequences. The choice of the assessment methodology is not just a technical matter; it can have political implications.

Rothwell and Zegveld (1985) specify distinct stages of policy development and implementation. Any evaluation of policy needs to consider these. The process starts with the establishment of national goals based on the assessment of economic, political and social problems. Policy formulation is the development of these goals in conjunction with experts and interested parties. The third phase, the selection of policy tools, e.g. tax allowances or training grants, is the precursor to implementation. The last phase is a policy impact assessment which can include adjustments in the light of actual experience and a policy audit.

They make a further point that policy evaluation should look at the institutions involved in policy formulation and implementation, their procedures and their interrelationships. In their view this would indicate not only what changes occurred as a result of policy, but also how and why.

In the end, evaluation must be in terms of agreed criteria and indicators. Any judgement of success would consider, typically, the following:

- the utility of policy tools from the point of view of users;
- the extent of the take up of the support provided;
- the pattern of take-up, e.g. small and large companies, type of industry etc.;
- the qualitative changes induced, such as changes in attitudes.

16.16.2 The role of policy feedback

The risks and costs involved with major innovation schemes put a premium on learning systems. This is helped by feedback and systematic data collection. Where radical new policies and major funding requirements are considered, there is a case for an experimental policy system on the lines of the US Department of Commerce's Experimental Technology Incentive program (ETIP). Between 1972 and 1982 the then US National Bureau of Standards carried out over 200 experiments and studies to better understand how government policies impact on private sector inventiveness and how to improve these impacts. Major components of its strategy were:

1. the selection of agencies known to be inclined towards change;
2. collaborative problem solving by all stakeholders in the planning, implementing and evaluating the experiment;
3. acting as a neutral, third-party facilitator;
4. continuing feedback of data on the experiment's impact on the private sector;
5. emphasis on transferring ETIP's skills, perception and methods to the agencies and institutionalizing the changes found to be beneficial.

16.17 SUMMARY

A government which wishes to generate a competitive, technology-based economy needs to establish coherent objectives and strategies. These must be integrated with its other policies. It must stimulate industry, facilitate technological progress and support longer-term research and technology development programmes.

An organizational framework is required for the development and implementation of a national technology policy. A technological infrastructure, including education and training, has to be maintained and evolved to support future requirements. The encouragement of inward technology transfer, the public procurement of advanced technology products and systems, continued participation in fundamental research and structured research intelligence all support a national technological capability.

The formulation of a technology policy is challenged by the conflicting value patterns of a pluralistic society, limited systematic data and problems of technology forecasting. Policy audits provide feedback and learning opportunities.

FURTHER READING

Bangemann, M. (1992) *Meeting the Global Challenge*, Kogan Page, London.

Dosi, G. (1984) *Technical Change and Industrial Transformation*, The Macmillan Press Ltd, London.

Kodama, F. (1991) *Analyzing Japanese High Technologies*, Pinter Publishers, London.

Kuehn, T.J. and Porter, A.L. (eds) (1981), *Science, Technology and National Policy*, Cornell University Press, Ithaca, New York.

McMillan, C.J. (1985) *The Japanese Industrial System*, (2nd edn), Walter de Gruyter, Berlin.

OECD, (1991) *Technology in a Changing World, (The Technology/Economy Program)*, OECD, Paris.

Porter, M.E. (1985) *The Competitive Advantage of Nations*, The Macmillan Press Ltd, London.

Rothwell, R. and Zegfeld, W. (1985) *Reindustrialization and Technology*, Longman Goup Ltd, Harlow.

Tassey, Gregory (1992) *Technology Infrastructure and Competitive Position*, Kluwer Academic Publishers, Norwell, MA.

Tisdell, C.A. (1981) *Science and Technology Policy*, Chapman & Hall, London.

United Nations, (1979) *Technology Assessment for Development*, United Nations, New York.

Appendix A
Formulae for
investment appraisal

A.1 THE PAYBACK METHOD

Let E = the project capital expenditure

R_i = the project benefit in the 'ith' period

then payback has been achieved when

$$E = \sum_{i=1}^{t} R_i$$

the value of 't' which solves *the equation* will be the payback period.

A.2 THE RETURN ON INVESTMENT ANALYSIS

Let r_i = rate of return for period i, and

$$r_i = \frac{R_i}{E} \ (100\%)$$

For this analysis 'R_i' is the project benefit after depreciation.

Where the returns vary over the product or process life 'T' we obtain an average rate of return r_a, where

$$r_a = \frac{100}{T} \left(\sum_{i=1}^{T} r_i \right) \% = \frac{100}{T} \left(\sum_{i=1}^{T} \frac{R_i}{E} \right) \%$$

A.3 THE DISCOUNTED CASH FLOW METHODS

A.3.1 The yield or rate of return method

This method obtains the rate of return for a project when the time value of money is taken into account. The rate of return here is based on the discounted values of future project benefits.

Let r = rate of return to be established.

Then with the previous notation we solve the equation

$$\sum_{i=1}^{T} \frac{R_i}{(1+r)^i} - E = 0$$

This means that the discounted values of all project benefits equal the initial project cost and a rate of return 'r' on the outstanding investment is obtained throughout the project life.

A.3.2 The net present value (NPV) method

Here, a target rate of return 's' is set by the firm and the NPV is determined. The NPV is given by the expression:

$$\sum_{i=1}^{T} \frac{R_i}{(1+s)^i} - E$$

If NPV > 0 the actual rate exceeds the target rate of return;
if NPV = 0 the actual rate equals the target rate;
if NPV < 0 the actual rate is less than the target rate.

Appendix B
An example of stochastic investment evaluation

Consider a new product development project which is evaluated by the discounted cash flow method on the basis of six different sets of assumptions. Using the corporate target rate of return, the following net present values (NPVs) are obtained with probabilities as stated:

NPV (£000)	Probability
1515	0.1
4402	0.2
5000	0.2
8259	0.2
4419	0.2
9525	0.1

The mean project NPV (u_m) can then be obtained from the equation

$$u_m = \sum_{i=1}^{n} u_i p_i$$

where u_i = the NPV of the 'ith' cash flow profile
p_i = the probability of the 'ith' cash flow profile
The mean project NPV for this set of data is £5 520 000
   The measure of risk is obtained from the project coefficient of variation 'C' where

$$C = \frac{\sqrt{\text{variance}}}{\text{mean}} = \frac{\text{standard deviation}}{\text{mean}}$$

The variance 'V' is calculated from the relationship

$$V = \sum_{i=1}^{n} (u_i - u_m)^2 p_i$$

The standard deviation for this set of data is £2 291 000.
   The project coefficient of variation is then

$$C = \frac{2291000}{5520000} = 0.415$$

The larger the coefficient of variation, the greater will be the project risks. If the value of 'C' is uncomfortably high, it might be better to investigate the development proposal further to reduce the level of risk.

References

Abernathy, W.J. (1978) *The Productivity Dilemma*, The Johns Hopkins University Press, Baltimore.

Acme (1991), *Manufacturing Research Strategy for the Nineties*, Science and Engineering Research Council (SERC), Swindon.

Advisory Council on Science and Technology (1991) *Science and Technology Issues*, HMSO, London.

Alvey Directorate (1988), *Alvey Programme*, Annual Report, Institution of Electrical Engineers, London.

Amendola, M. and Gaffard, J-L (1988) *The Innovative Choice*, Basil Blackwell, Oxford.

Andersen Consulting, Cardiff Business School and Cambridge University (1992) *Lean Enterprise Benchmarking Project*, Andersen Consulting, London.

Andrews, David (1971) Reduction of development cost uncertainty. *Management Accounting*, September to November 1971.

Ansoff, I.H. (1979) *Strategic Management*, Macmillan Publishers, London.

Ansoff, I.H. and Stewart, J.M. (1967) Strategies for a technology-based business. *Harvard Business Review*, **45**(6).

Bangemann, M. (1992) *Meeting the Global Challenge*, Kogan Page, London.

Banker, R.D. *et al.* (1990) Cost of product and process complexity, in *Measures for Manufacturing Excellence*, (ed. R.S. Kaplan) Harvard Business School Press, Boston, MA.

Batty, J. (1988) *Accounting for Research and Development*, Gower Publishing Co Ltd, Aldershot.

Berg, M.R. (1981) The politics of technology assessment, *Science, Technology, and National Policy*, (eds T.J. Kuehn and A.L. Porter) Cornell University Press, Ithaca, New York.

Bessant, J. and Grunt, M. (1985) *Management and Manufacturing Innovation in the United Kingdom and West Germany*, Gower, Aldershot.

Burgelman, R.A. and Maidique, M.A. (1988) *Strategic Management of Technology and Innovation*, Irwin, Homewood, IL.

Buchanan, D.A. and Boddy, D. (1983) *Organizations in the Computer Age*, Gower, Aldershot.

Busby, J.S. (1992) *The Value of Advanced Manufacturing Technology: How to Assess the Worth of Computers in Industry*, Butterworth Heinemann, Oxford.

Calori, R. and Noel, R. (1986) Successful strategies in French high technology companies. *Long Range Planning*, **19**(6), 54–65.

Carter, L.R. and Lowe, P.H. (1974) Evolutionary operation: an aid for better plant performance. *The Production Engineer*, **53**(1).

Central Statistical Office (1979) *Standard Industrial Classification*, Revised 1980, HMSO, London.

Cetron, Marvin, J. (1969) *Technological Forecasting*, Gordon and Breach Science Publishers, New York.

References

Chandler, A.O. (1962) *Strategy and Structure*, MIT Press, Cambridge, MA.
Chew, W.B. *et al.* (1990) Measurement, coordination and learning in a multiplant network, in *Measures for Manufacturing Excellence*, (ed. R.S. Kaplan) Harvard Business School Press, Boston, MA.
Child, J. (1984) *Organization: A Guide to Problems and Practice*, Harper and Row, London.
Clark, J. *et al.* (1988) *The Process of Technological Change*, Cambridge University Press, Cambridge.
Cobb, C.W. and Douglas, P.H. (1928) A theory of production. *American Economic Review*, **XVIII** (139).
Commission of the European Communities (1991) *Science and Technology in Europe*, Office for Official Publications of the European Communities, L-2985, Luxembourg.
Coombs, R. *et al.* (1987) *Economics and Technological Change*, Macmillan Education, Basingstoke.
Coombs, R., Saviotti, P. and Walsh, V. (eds) (1992) *Technological Change and Company Strategies*, Academic Press Ltd, London.
Cooper, R. and Turney, P. (1990) Internally focused activity based cost systems, in *Measures for Manufacturing Excellence*, (ed. R.S. Kaplan) Harvard Business School Press, Boston, Ma.
Corlett, E.N. and Richardson, J. (eds) (1981) *Stress, Work Design and Productivity*, John Wiley & Sons, Chichester.
Cornwall, J. (1977) *Modern Capitalism: Its Growth and Transformation*, Martin Robertson, London.
Cox, T. (1978) *Stress*, The Macmillan Press Ltd, London.
Cummings, T. and Blumberg, M. (1987) Advanced manufacturing technology and work design, in *The Human Side of Advanced Manufacturing Technology*, (eds T.D. Wall *et al.*) John Wiley & Sons, Chichester.
Cyert, R.M. and March, J.G. (1963) *A Behavioural Theory of the Firm*, Prentice-Hall, New York.
Danila, N.V. (1989) Strategic formulation of high technology projects using the support graph, *Technology Analysis and Strategic Management*, **1**(3) 273.
De Garmo, P.E. *et al.* (1979) *Engineering Economy*, 2nd edn, Macmillan Publishing Inc., New York.
De Greene, K.B. (1982) *The Adaptive Organization: Anticipation and Management of Crises*, Wiley, New York.
De Leon, P. (1976) *A Cross-National Comparison of Nuclear Development Strategies*, Rand Corporation, Santa Monica, California.
Denison, D.R. (1990) *Corporate Culture and Organizational Effectiveness*, John Wiley & Sons, New York.
Department of Trade and Industry, (1973) *Rolls-Royce Limited, Investigation under Section 165 (a) (i) of the Companies Act 1948*, HMSO, London.
Dicken, P. (1986) *Global Shift: Industrial Change in a Turbulent World*, Harper & Row, London.
Dodgson, M. (ed) (1989) *Technology Strategy and the Firm: Management and Public Policy*, Longman and Science Policy Research Unit, Brighton.
Dosi, G. (1984) *Technical Change and Industrial Transformation*, The Macmillan Press Ltd, London.
DRT International, (1991) *Research and Development Cost Sharing: A Guide for Multinational Enterprises*, DRT International, New York.
Dussauge, P., Hart, S. and Ramanantsoa, B. (1992) *Strategic Technology Management*, John Wiley & Sons, Chichester.
Edge, G. (1990) Investment in technology, in *Technology and Management*, (ed. R. Wild) Cassell Educational Ltd, London.

Edosomwam, J.A. (1989) *Integrating Innovation and Technology Management*, John Wiley & Sons, New York.

Engel, A.K. (1987) The Japan 'creative science' projects. *Proceedings of International Conference on Japanese Information in Science, Technology and Commerce.* University of Warwick, September 1987. Eds S.V. King and G.J. Sassoon, British Library Japanese Information Service, London.

Engelsman, E.C. and van Raan, A.F. (1991) *Mapping of Technology. A first exploration of knowledge diffusion amongst fields of technology.* Policy Studies on Technology and Economy. Ministry of Economic Affairs, The Hague, The Netherlands.

Esch, M.E. (1968) Planning assistance through technical evaluation of relevance numbers. *Second Annual Technology Management Conference on Technological Forecasting and Planning.* Lake Placid, New York.

European Foundation for the Improvement of Living and Working Conditions (1992) *Roads to Participation in the European Community – Increasing Prospects of Employee Representation in Technical Change.* Report by D. Frölich; C. Gill and H. Kreiger, EFILWC, Dublin, Ireland.

Farrell, J. and Saloner, G. (1987) Competition, compatibility and standards, in *Product Standardization and Competitive Strategy*, (ed. L. Gabel) North Holland Publishing Co, Amsterdam.

Ford, J.D. (1988) Develop your technology strategy. *Long Range Planning*, **21**(5), 85–95.

Freeman, C. (1987) *Technology Policy and Economic Performance*, Pinter Publishers, London.

Freeman, C. and Perez, C. (1988) Structural crises of adjustment: business cycles and investment behaviour, in *Technical Change and Economic Theory*, (eds G. Dosi *et al.*) Pinter, London.

Frieman, W. (1987) Beyond translation: using technical literature for competitive assessments. *Proceedings of International Conference on Japanese Information in Science, Technology and Commerce.* University of Warwick, September 1987. Eds S.V. King and G.J. Sassoon, British Library Japanese Information Service, London.

Funk, J.L. (1993) Japanese product development strategies. *IEEE Transactions on Engineering Management*, **40**(3), 224.

Gabel, H.L. (ed) (1987) *Product Standardization and Competitive Strategy*, North Holland Publishing Co., Amsterdam.

Gibbons, M. (1982) The evaluation of government policies for innovation. Paper presented to *Six Countries Programme Workshop*, Windsor.

Girifalco, L.A. (1991) *Dynamics of Technological Change*, Van Nostrand Reinhold, New York.

Glasser, A. (1982) *Research and Development Management*, Prentice-Hall Inc, New York.

Harrison, F.L. (1992) *Advanced Project Management*, (3rd edn,) Gower Publishing Co Ltd., Aldershot.

Harvey-Jones, J. (1993) *Managing to Survive*, William Heinemann Ltd, London.

Herzberg, F.W. (1959) *The Motivation to Work*, John Wiley & Sons, New York.

House of Commons (Session 1977–78) *Second Report from the Select Committee on Science and Technology, Innovation Research and Development in Japanese Science Based Industry*, HC682-I, HMSO, London.

Höusser, E. (1981) Patent information – an important factor for the industrial development of the market economy countries. *Conference proceedings 'The Role of Patent Information in the Transfer of Technology'.* Ed. F.A. Sviridov, Pergamon International Information Corporation, McLean, VA

Imai, M. (1986) *Kaizen: The Key to Japan's Competitive Success*, McGraw-Hill Publishing Company, New York.

IMR Ltd (1992) *Key Issues and Future Directions in Research and Development for High Technology Companies*, KMPG Peat Marwick High Technology Practice, London.

Institution of Electrical Engineers (1993) *A Report on an OSTEMS Visit to Japan*, IEE, London.

Jacobsson, S. (1986) *Electronics and Industrial Policy: The Case of Computer Controlled Lathes*, Allen & Unwin, London.

Jantsch, E. (1967) *Technological Forecasting in Perspective*, OECD, Paris.

Johnson, G.N. (1988) Rethinking incrementalism. *Strategic Management Journal*, **9** 75–91.

Johnson, H. Thomas, (1990) Performance measurement for competitive excellence, in *Measures for Manufacturing Excellence* (ed. R.S. Kaplan) Harvard Business School Press, Boston, MA.

Jones, D.T. (1989) Corporate strategy and technology in the world automobile industry, in *Technology Strategy and the Firm: Management and Public Policy*, (ed. M. Dodgson) Longman and Science Policy Research Unit, Brighton.

Jones, D.T. (1992) Beyond the Toyota production system: the era of lean production, in *Manufacturing Strategy* (ed. C.A. Voss) Chapman & Hall, London.

Kanter, R.M. (1983) *The Change Masters*, Simon & Schuster Inc, New York.

Kaplan, R.S. (ed) (1990) *Measures for Manufacturing Excellence*, Harvard Business School Press, Boston, MA.

Kodama, F. (1991) *Analyzing Japanese High Technologies*, Pinter Publishers, London.

Kuhn, T.S. (1962) *The Structure of Scientific Revolutions*, University of Chicago Press, Chicago, IL.

Langrish, J. *et al.* (1971) *Wealth from Knowledge*, The Macmillan Press Ltd, London

Lawrenson, P.J. (1993) Know-how as product; creation, exploitation and national benefit. *Engineering Management Journal*, 3(1) February.

Lazarus, R.S. (1976) *Patterns of Adjustment*, McGraw-Hill, New York.

Lea, R.H. and Miller, P.S. (1968) The Multiple contingency concept of long-range technological planning, in *Technological Forecasting for Industry and Government*, (ed. J.R. Bright) Prentice-Hall Inc, Englewood Cliffs, NJ.

Levitt, B. and March, J.G. (1988) Organizational learning. *Annual Review of Sociology*, 319–340.

Linstone, H.A. and Turoff, M. (eds) (1975) *The Delphi Method – Techniques and Applications*, Addison-Wesley Publishing Company, Reading, MA.

Little, A.D. (1981) *The Strategic Management of Technology*, Arthur D. Little, New York.

Lloyd, M.R. and Mason, M. (1993) Large electrical machines: powering a business renaissance for the twenty-first century. *Proceedings Part B, Institution of Mechanical Engineers*, **207**.

Lock, D. (1992) Project Management, Gower Publishing Company Ltd, Aldershot.

Loveridge, R. and Pitt, M. (eds) (1990) *The Strategic Management of Technological Innovation*, John Wiley & Sons Ltd, Chichester.

Lowe, P. (1979) *Investment for Production*, Associated Business Press, London.

Lupton, T. (1965) The practical analysis of change in organization, *Journal of Management Studies*, May.

Lyon, R.H. and Rydz, J.S. (1981) Technology assessment in a consumer product manufacturing company, Management Dvision. *ASME*, 81-WA/MGT-1.

Maidique, M.A. and Patch, P. (1978) Corporate strategy and technological policy. *Harvard Business Review*, May–June.

Malpas, R. (1987) Strategies for success in manufacturing. Proceedings, Institution of Mechanical Engineers, **201**(B3) 143.

Mansfield, E. (1961) Technical change and the rate of imitation. *Econometrica*, **29**(741).

Mansfield, E. (1969) *Industrial Research and Technological Innovation*, Longmans, Green & Co. Ltd, London.

Mansfield, E. *et al.* (1972) *Research and Innovation in the Modern Corporation*, The Macmillan Press Ltd, London.

Mansfield, E. *et al.* (1981) Imitation costs and patents: an empirical study. *Economic Journal*, **91**, 907–81.

Maslow, A.H. (1943) A theory of human motivation. *Psychological Review*, **50**.

McLoughlin, I. and Clark, J. (1988) *Technological Change at Work*, Open University Press, Milton Keynes.

McMillan, C.J. (1985) *The Japanese Industrial System*, 2nd edn, Walter de Gruyter, Berlin.

Mitchell, G.R. (1988) Options for the strategic management of technology. *Proceedings of the First International Conference on Technology Management*, Interscience Enterprises Ltd, Geneva, Switzerland.

Moenaert, R. *et al.* (1990) Turnaround strategies for strategic business units with an ageing technology, in *The Strategic Management of Technological Innovation*, (eds R. Loveridge and M. Pitt) John Wiley & Sons Ltd, Chichester.

Nabseth, L. and Gray, G.F. (1974) *The Diffusion of New Industrial Processes*, Cambridge University Press, Cambridge.

National Economic Development Office (NEDO) (1985) *Guidelines for the Implementation of Advanced Manufacturing Technology (AMT)*, NEDO, London.

National Economic Development Council (NEDC) (1989) *Technology Transfer Mechanisms in the UK and Leading Competitor Nations*, NEDC, London.

Nelson, R. (1974) Intellectualizing the moon-ghetto metaphor: a study of the current malaise of rational analysis in social problems. *Policy Sciences*, **5**.

Nikkan Kogyo Shimbun Ltd (ed.) (1988) *Poka-yoke. Improving Product Quality by Preventing Defects*, Productivity Press, Cambridge, MA.

OECD (1991) *Technology in a Changing World (The Technology/Economy Program)*, OECD, Paris.

OECD (1993) *The Measurement of Scientific and Technical Activities: Proposed Standard Practice for Surveys of Research and Experimental Development*. Frascati Manual, 5th edn, Organization for Economic Cooperation and Development, Paris.

Osola, V.J. (1986) The initiation and management of change. *Chartered Mechanical Engineer*, December, 22.

Ozawa, T. (1974) *Japan's Technological Challenge to the West 1950–1974*, MIT Press, Cambridge, MA.

PA Consulting Group (1989) *Attitudes to R&D and the Application of Technology*, PA Consulting Group, London.

Parker, R.C. (1982) Increasing employment. *Chartered Mechanical Engineer* (CME), November.

Pavitt, K. (1984) Sectorial patterns of technical change: towards a taxonomy and a theory. *Research Policy*, **13**, 343–373.

Pettigrew, A.M. (1977) Strategy formulation as a political process. *International Studies of Management and Organization*, **7**(2), 78–87.

Polanyi, M. (1958) *Personal Knowledge*, Routledge & Kegan Paul, London.

Porter, M.E. (1985) *Competitive Advantage*, Collier Macmillan Publishers, London.

Porter, M.E. (1990) *The Competitive Advantage of Nations*, The Macmillan Press Ltd, London.

Putnam, Hayes & Bartlett, Inc, (1982) *The Impact of Private Voluntary Standards on Industrial Innovation*, Report to the National Bureau of Standards, US Department of Commerce, NBS GCR 82-420.

Ramaer, J.C. (1979) The choice of appropriate technology by a multinational corporation: a case study of Messrs Philips, Eindhoven, in *Appropriate Technologies for Third World Development*, ed. A. Robinson, The Macmillan Press, London.

Ranganthan, S.R. (1955) Prolegomena to Library Classification, Advent Books Inc., New York.

Reddy, N.M. (1987) Voluntary product standards: linking technical criteria to marketing decisions. IEEE Transaction on Engineering Management, **Vol EM-3** (4) November.

Revelle, R. (1975) The scientist and the politician. *Science*, **187**, 1100.

Rosenberg, N. (1982) *Inside the Black Box*, Cambridge University Press, Cambridge.

Roth, A.V. *et al.* (1992) Operating strategies for the 1990s: elements comprising world-class manufacturing, in *Manufacturing Strategy*, (ed. C.Voss) Chapman & Hall, London.

Rothwell, R. and Zegveld, W. (1985) *Reindustrialization and Technology*, Longman Group Ltd, Harlow.

Rouse, W.B. (1992) *Strategies for Innovation*, John Wiley & Sons Inc, New York.

Roussell, P.A., Saad, K.N. and Erickson, T.J. (1991) *Third Generation R&D*, Harvard Business School Press, Boston, MA.

Ryan, C.G. (1984) *The Marketing of Technology*, Peter Peregrinus Ltd, London.

Schumacher, E.F. (1973) *Small is Beautiful*, Blond & Briggs Ltd, London.

Schumpeter, J.A. (1961) *The Theory of Economic Development*, Oxford University Press, Oxford (First published in German in 1911).

Shingo, S. (1985) *A Revolution in Manufacturing, the SMED System*, Productivity Press, Cambridge, MA.

Solow, R.M. (1957) Technical change and the aggregate production function. *The Review of Economics and Statistics*, **39**, 312.

Stoneman, P. (1983) *The Economic Analysis of Technological Change*, Oxford University Press, Oxford.

Tassey, G. (1986) The role of the National Bureau of Standards in supporting industrial innovation. *IEEE Transactions on Engineering Management*, **Vol EM-33**, 3.

Tassey, G. (1992) *Technology Infrastructure and Competitive Position*, Kluwer Academic Publishers, Norwell, MA.

Terborgh, G. (1949) *Dynamic Equipment Policy*, McGraw-Hill, New York.

Terborgh, G. (1958) *Business Investment Policy*, Machinery and Allied Products Institute, Chicago, IL.

Tidd, J. (1991) *Flexible Manufacturing Technology and International Competitiveness*, Pinter Publishers, London.

Tischler, A.O. (1969) A commentary on low-cost space transportation. *Astronautics and Aeronautics*, **7**(8) August 50–64.

Tranfield, D. and Smith, S. (1990) *Managing Change*, IFS Publications, Bedford.

Turner, F. (1993) Business systems engineering, *Proceedings of the Institution of Mechanical Engineers*, **207**.

Twiss, B. (1986) *Managing Technological Innovation*, 3rd edn, Pitman Publishing, London.

Twiss, B. (1990) Business strategies for new technologies, in *Technology and Management*, (ed. R. Wild) Cassell Educational Ltd, London.

Twiss, B. and Goodridge, M. (1989) *Managing Technology for Competitive Advantage*, Pitman Publishing, London.

Urban, G.L. and Hauser, J.R. (1980) *Design and Marketing of New Products*, Prentice-Hall, Englewood Cliffs, NJ.

US Department of Commerce, Technology Administration, (1990) *Emerging Technologies. A Survey of Technical and Economic Opportunities*, Washington, DC.

Veldhuis, K.H. (1979) Transfer and adaptation of technology: Unilever as a case study, in *Appropriate Technologies for Third World Development*, (ed. A. Robinson) The Macmillan Press, London.

Vickery, B.C. (1975) *Classification and Indexing in Science*, 3rd edn, Butterworth, London.

von Hippel, E. (1990) *The Sources of Innovation*, Oxford University Press, New York.

Warnecke, H.J. and Schraft, F.W. (1982) *Industrial Robotics: Application Experience*, IFS Publications Ltd, Bedford.

Warren, K. (1992) The politics of industry. *Proceedings of the Institution of Mechanical Engineers*, **206**, 151.

White, G.C. (1982) *Technological Changes and Employment*, Work Research Unit, Occasional Paper 22, Department of Employment, London.

White, G.C. (1984a) *Managing Stress in Organizational Change*, Work Research Unit, Occasional Paper 31, Department of Employment, London.

White, G.C. (1984b) *Employee Involvement in Work Design*, Work Research Unit, Occasional Paper 29, Department of Employment, London.

Woodward, J. (1965) *Industrial Organization: Theory and Practice*, Oxford University Press, London.

Zimmerman, L.W. and Hart, G.D. (1982) *Value Engineering: A Practical Approach for Owners, Designers and Contractors*, Van Nostrand Reinhold Company, New York.

Index

Page numbers appearing in **bold** refer to figures and page numbers appearing in *italic* refer to tables.